THE COMPLETE
GARDEN BIRD
BOOK

GARDEN BIRD
BOOK

HOW TO IDENTIFY AND ATTRACT BIRDS TO YOUR GARDEN

Text by MARK GOLLEY *with* STEPHEN MOSS
Illustrated by DAVID DALY

NEW
HOLLAND

First published in 1996 by
New Holland (Publishers) Ltd
London • Cape Town • Sydney • Singapore

Reprinted 1996

24 Nutford Place
London W1H 6DQ
United Kingdom

80 McKenzie Street
Cape Town 8001
South Africa

3/2 Aquatic Drive
Frenchs Forest, NSW 2086
Australia

ISBN 1 85368 580 1 (hbk)
ISBN 1 85368 581 X (pbk)

Commissioning Editor: Jo Hemmings
Editors: David Christie and Sophie Bessemer
Designer: Alan Marshall, Wilderness Design, Kent

Typeset by Alan Marshall, Wilderness Design, Kent
Reproduction by Hirt and Carter, Cape Town, South Africa
Printed and bound in Malaysia by Times Offset (M) Sdn. Bdh.

There could be a small book written in itself to cover all the people I'd like to thank, for all
manner of reasons, with regard to this project, but sadly I only have a handful of paragraphs.

Firstly, a big thank you to Dave Daly for providing such truly excellent art-work. His
professionalism and talent shine through.

I would also like to thank Jo Hemmings and Sophie Bessemer at New Holland (Publishers)
for being such an enormous help.

A particular note of thanks must go to two birdwatchers from Devon. To Gordon Vaughan
and Wally Towler who, many years ago, gave me all kinds of advice and help whilst I pursued
my favourite hobby as an itinerant schoolboy birdwatcher!

Many thanks to NY, TFC and MS for providing the perfect soundtrack which accompanied me
throughout. I'd have been lost without it.

Finally, my special thanks to Caroline for all her patience, encouragement and support over
the past months, and to my parents for all of these and their constant belief and backing in all
that I have done. This book is dedicated to them . . .

MARK GOLLEY

I would also like to thank Derek Toomer from the BTO Garden BirdWatch survey and Mike
Everett of the RSPB for their help in supplying invaluable information.

STEPHEN MOSS

CONTENTS

INTRODUCTION

WHETHER YOU LIVE IN THE HEART OF THE COUNTRY or the middle of a city, birds will be attracted to your garden. Indeed, one of the greatest pleasures of owning a garden is watching the birds that visit it.

If you enjoy looking at garden birds, you're not alone. In recent years, millions of people have discovered the joys of observing bird behaviour at close range. But it's not only humans who benefit – gardens are vital for the well-being and survival of birds, too.

Britain's gardens cover more than quarter of a million hectares, making them one of our most valuable bird habitats. In spring and summer, trees, bushes and artificial nest sites provide a range of places for birds to nest and rear their young. In winter, your garden can become a vital refuge for birds, especially if you regularly supply food and water. During prolonged cold spells, this can make the difference between life and death, particularly for smaller species.

This book has two principal aims: to show you how to encourage birds to visit your garden, by providing places for them to feed, drink and nest; and to enable you to identify the different species.

Section One (Watching and Attracting Birds to your Garden) contains practical advice on how to create the best habitats for birds, and on providing food, water and nest sites. It also tells you which species you can expect to visit your garden at different times of the year.

Section Two (Identifying the Birds) consists of seventy fully-illustrated, double-page spreads, each covering a single species. These will help you identify the visitors to your garden, as well as informing you about their habits. Each spread includes illustrations of the bird in various different plumages, showing different forms of behaviour.

Finally, there is a short section, ***Useful Addresses and Further Reading***, *(see page 174)* to help you make the most of your interest in garden birds.

BIRD TOPOGRAPHY

BIRD TOPOGRAPHY, AS ITS NAME SUGGESTS, is simply a description of the external features of a bird. Without a clear agreement amongst birdwatchers about what each particular part of a bird is called, precise identification would be much more difficult.

Some terms used in bird topography are self-explanatory: for example the bill, legs and crown. Others are specialised terms, often baffling to the beginner, such as tertials, lores and supercilium.

In fact, with practice, learning these terms is not so hard as it first appears. Bird feathering follows a logical pattern and many topographical terms reflect this. For example, the pattern of feathers on a bird's wing is more or less the same whatever the species.

The garden is an ideal place to learn about bird topography, as you normally get close and prolonged views. Try comparing the birds you see with the diagram, and see if you can spot the different areas. You will soon find that it adds a new dimension to your appreciation and knowledge of garden birds.

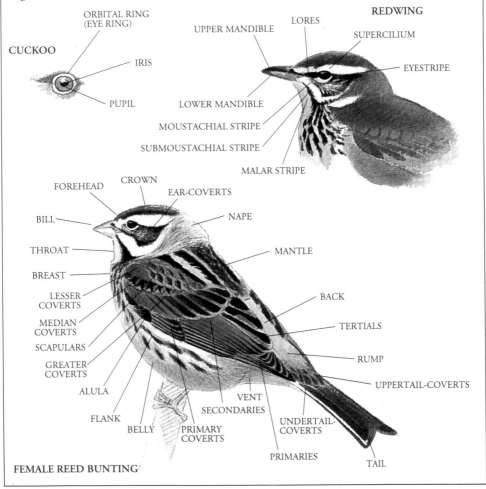

ORBITAL RING (EYE RING)
UPPER MANDIBLE
LORES
REDWING
SUPERCILIUM
CUCKOO
IRIS
EYESTRIPE
PUPIL
LOWER MANDIBLE
MOUSTACHIAL STRIPE
SUBMOUSTACHIAL STRIPE
MALAR STRIPE

FOREHEAD
CROWN
EAR-COVERTS
NAPE
BILL
THROAT
MANTLE
BREAST
LESSER COVERTS
BACK
MEDIAN COVERTS
TERTIALS
SCAPULARS
GREATER COVERTS
RUMP
ALULA
UPPERTAIL-COVERTS
VENT
FLANK
SECONDARIES
BELLY
PRIMARY COVERTS
UNDERTAIL-COVERTS
PRIMARIES
TAIL

FEMALE REED BUNTING

The sturdy, hooked bill of the Sparrowhawk is ideally suited to tearing apart the flesh from small mammals and birds, leaving little wastage.

The compact, neat bill of the Chiffchaff allows plenty of opportunity for catching small insects such as greenfly. Crumbs and suet from a bird table are also easy 'prey'.

The Bullfinch has a broad, sharp-edged bill, ideal for stripping buds and crushing seeds and fruits.

Blackbirds, like many birds common to town and country, have an 'all-purpose' bill allowing them to eat a wide variety of foods.

WATCHING *and* ATTRACTING BIRDS *to* YOUR GARDEN

WATCHING
GARDEN BIRDS

L OOK OUT OF THE WINDOW INTO YOUR GARDEN, and the chances are you'll see some birds. They may be regular visitors, such as Blue Tit, Blackbird or Robin. Or they may be unfamiliar, and tricky to identify.

So, where do you go from here? Well, the first step is to get a reliable pair of binoculars (see page 14). Binoculars allow you to observe the birds, without being noticed, at close quarters, opening up a whole new world of interest.

It is also worth keeping a log of the birds which visit your garden. Note down the different species you see, as well as a record of the date, time of day, and how many birds are present. This will help you find out which species regularly come into your garden, and which are more casual visitors.

If you do see something unusual, jot down details of the bird: especially its size, plumage details and behaviour. Birds may stay for only a moment or two before flying off, so it is important to note down as much detail as possible. Afterwards you can use **Section Two** of this book to identify the bird at your leisure.

This Song Thrush is using the 'anvil technique' to eat a snail. The bird smashes the snail's shell repeatedly on a stone, rock or other hard surface in order to get at its soft contents. Look for tell-tale signs of clusters of smashed snail shells, often on the edge of a garden path.

If you have a pond in your garden, you can expect a visit from a Grey Heron. Herons feed mainly at dawn and dusk, when the glare from the sun is at its lowest level, enabling them to see their prey more easily. However, if you value your fish, you should cover your pond with fine netting, or it will soon be empty!

10

BTO Garden BirdWatch survey

The British Trust for Ornithology's Garden BirdWatch is the largest year-round survey of birds in the British Isles. More than three thousand people regularly collect data on the birds visiting their gardens, which are then analysed by the BTO to produce a 'top ten' of the most regular garden birds, and a wealth of other information.

The aim of the Garden BirdWatch is to monitor the population changes of some of our most familiar species, providing an early-warning system of potential declines, as well as evidence of rises in population. Along with the information from other BTO surveys, such as the Common Birds Census and the Breeding Birds Survey, the Garden BirdWatch enables conservationists to preserve and protect our birdlife.

If you would like to participate in the Garden BirdWatch, contact the BTO (see page 174). There is a registration fee of £10 to cover administrative costs, but, in return, participants get discounts on bird food and receive a quarterly magazine containing articles and the latest survey results.

RESULTS OF GARDEN BIRD SURVEY, 1992-1994: species most frequently recorded.		
	SPECIES	RISE OR DECLINE?
1	Blackbird	Small rise
2	Blue Tit	Small rise
3	House Sparrow	Small decline
4	Robin	Rise
5	Starling	Small decline
6	Dunnock	Rise
7	Great Tit	Rise
8	Chaffinch	Rise
9	Collared Dove	Rise
10	Greenfinch	Rise

The frequency with which different species are recorded varies from season to season, with Blue Tit slightly more frequent than Blackbird during the autumn and winter months. However, when data for the whole year round are analysed, the Blackbird emerges as the species most often present in Britain's gardens.

The survey, which recorded more than fifty species in all, has also revealed fascinating details about our less familiar garden birds. One of the most unexpected results was the huge rise in the numbers of Siskins visiting gardens. In the very first garden bird survey, more than a quarter of a century ago, this attractive finch was recorded in only 7% of gardens. By 1994 this figure had risen to an amazing 69%.

Siskins have enjoyed a population explosion in recent years, thanks to the rise in conifer plantations, which provide food and nest sites. But they also seem to have changed their habits, being attracted to gardens by the availability of nuts in bird feeders. The survey found peak numbers in gardens during March and April, but Siskins are also seen throughout the rest of the year in some localities, particularly where there are suitable woods for nesting nearby.

The term 'bird behaviour' refers to all the ways birds live their lives, from feeding to roosting and birdsong to courtship displays. Many types of behaviour occur on a daily basis, though some, such as migration or courtship, are seasonal.

A garden is the perfect place to study bird behaviour at close range. Whether watching tits feeding on a nut-basket or a Blackbird sunbathing, you are witnessing one aspect or another of the way birds behave.

Once you have begun to identify the different birds in your garden, your next step is to find out more about the way they behave. At first this can seem a daunting task, but with a little effort you will soon begin to understand why birds behave in a particular way.

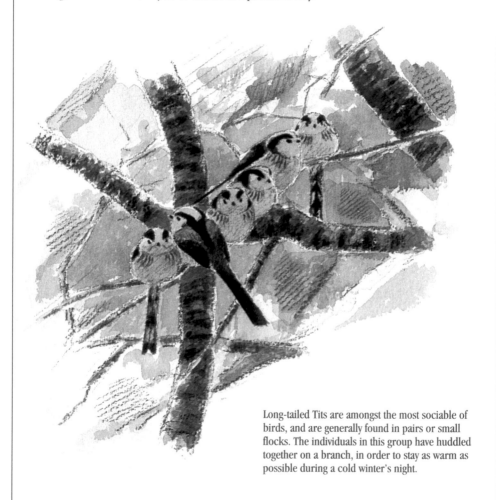

Long-tailed Tits are amongst the most sociable of birds, and are generally found in pairs or small flocks. The individuals in this group have huddled together on a branch, in order to stay as warm as possible during a cold winter's night.

▶ Birdsong has two functions: to defend a breeding territory against rival males, and to attract a mate. Like all songbirds, Robins are early risers, singing most during the hours around dawn. However, they can often be heard later in the day as well.

◀ During the spring, look out for birds carrying nesting material in their beaks. Spotted Flycatchers often build their nest in a hollow in a tree or a crack in a wall. They use bits of grass and small twigs, and line the nest with feathers.

▼ Birds' feathers get dirty very easily, and need frequent cleaning to keep them in good condition. This House Sparrow is using a birdbath to wash the city grime from its plumage. If you have a birdbath, make sure you change the water frequently to prevent it getting too dirty.

A QUICK GLANCE THROUGH ANY OF THE REGULAR MONTHLY BIRD MAGAZINES will leave you reeling at the vast range of binoculars that are now available. Obviously, if you are interested in just looking at birds in your garden then it is perhaps a little extravagant to spend large amounts for a top-of-the-range pair of binoculars. There is, however, a wide variety of reasonably priced, lightweight, easy-to-use binoculars which are ideally suited to your needs.

But how do you decide which binoculars are for you? The ideal thing to do is to go to one of the many specialist shops that deal in optical equipment. Explain to staff what you are looking for, how much you are willing to pay and what you will use them for, and they will know what to suggest.

For birds in the garden, you do not need to choose a 'big' pair of binoculars. The magnification need only be around 8 x 30 – and these will give you clarity, a good, bright image, and they should be nice and light. Choose the binoculars you feel most comfortable with, the ones that feel 'right' in your hands, be they porro prism, roof prism or a more compact design.

Porro prism
The 'traditional'-looking binoculars in a style which, despite refinements, has not really changed for decades. Often, these binoculars can be heavy, bulky and a little impractical for use, especially if you are looking out of the kitchen window. The magnifications vary, from 7 x 50 upwards to 12 x 50 and beyond, but there are many porro prism binoculars that fall into the ideal category – usually 8 x 30 – and so certainly bear these in mind.

Compact binoculars
Perhaps the best option for simple enjoyment of birds in your garden is a pair of neat, very easy-to-use, very lightweight, 'put them in your pocket', compact binoculars. What they lack in magnification, often only up to 7 x 26, they make up for in the ways noted above. Some of the leading manufacturers of optics have now turned their eyes to this end of the market.

Roof prism
Often at the expensive end of the scale, roof-prism binoculars tend to be favoured by birdwatchers as they are easier to hold, with regard to both shape and weight. Not all roof-prism binoculars will have the bank manager seeing red, but, as you may expect, quality is not so good at the lower end of the market.

IDEAL GARDEN
HABITATS

WHENEVER POSSIBLE, IT IS ALWAYS WORTHWHILE trying to bring some diversity of habitat into your garden. If you have a lawn, pure and simple, what about putting aside some space to plant some shrubs or trees, or be brave and dig out a pond! If you already have a garden with trees in it, how about trying to plant a hedge along the outside of your garden?

Trees, hedges and shrubs are all very important habitats for garden birds. They provide natural cover, offer potential nesting places and are a tremendous food source. Of course there are hundreds of choices, but are there some that are better than others? Here, we investigate this.

If you have the resources, a pond will add another dimension to your garden and the birdlife in it. Ponds will quickly attract insects, on which birds will thrive, and of course you have a ready-made birdbath and drinking pool. Remember that water is as vital to birds as are worms, seeds, grain or nuts, and it should never be forgotten.

Blackberries come into their own in the autumn. The plump delicious fruit will be welcomed by both resident and migrant thrushes, all of which will have a field day feeding on the berries.

The **Wild Cherry** offers juicy fruit in the autumn. Bullfinches enjoy nibbling at the flesh, before tackling the hard stone inside. The beautiful flowers attract insects for pollination, which in turn attract birds – warblers particularly.

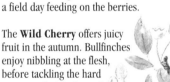

The **Birch** catkins of early spring are popular with some of the smaller finches, particularly Siskin and Redpolls. In winter when the tree loses its leaves, woodpeckers often become regular visitors to birch trees, finding the wood easy to penetrate for insects.

Both the **Norway Spruce** and the **Sitka Spruce** are popular with several species of bird, most notably Coal Tit, Goldcrest, Treecreeper, Nuthatch and finches.

Wild crab apples are an invaluable source of food for thrushes in hard weather. Redwings and Fieldfares will devour the flesh and leave the pips for Chaffinches and tits.

In spring and summer, a whole host of species ranging from woodpeckers to warblers can breed in **Oaks** with vast numbers of small caterpillars providing enormous amounts of food for parent birds to take back to hungry broods.

The **Hawthorn** is a 'dense' tree, offering protection for those species that choose to nest inside. The haws themselves are a food source for finches and thrushes and add a splash of colour to the garden.

The best way to attract birds to your garden is to provide as good a selection of plants as possible. Plants are valuable for several reasons. They provide shelter and cover, enabling birds to roost at night, and their thick foliage also enables nesting birds to protect their eggs and chicks against various predators.

Whatever the season, garden plants provide a ready source of food, including seeds, berries, fruit and nectar. One way to help birds survive the winter months is to plant a variety of berry-bearing shrubs in your garden.

Plants also play an important role by attracting insects, another vital source of food, especially for smaller birds such as tits and warblers. You should choose plants suitable for insects, such as Buddleia, often known as the butterfly bush because its flowers are so attractive to butterflies.

You can also attract birds to your garden by creating a pond. Even a small garden pond provides a place for birds to drink and bathe, as well as attracting a variety of insects. A pond is surprisingly simple to make, and can be built in even the smallest of gardens.

In hard winter weather, especially when frost or snow cover regular food supplies, windfall fruit is a vital source of food for birds. These Fieldfares and Redwings are winter visitors from Scandinavia.

Like all finches, the Goldfinch is a seed-eater, using its specially adapted beak to pluck individual seeds from the plant. Thistle, teazel and burdock seeds are favourites, although like this bird on the left, Goldfinches are also partial to dandelion seeds.

Nectar is another important source of energy, especially for smaller birds. This Blackcap is using its slender bill to probe into the Honeysuckle flower. As it does so, its face may become yellow with pollen.

Gardens are an important refuge for insects, especially if you avoid using powerful insecticides. Try to keep at least part of the garden in a 'wild' state, letting the native weeds take over. You may not be popular with your neighbours, but the insects – and the birds – will thank you for it! Bear in mind also that many of the plants that we tend to call 'weeds', are actually very attractive when in flower and can certainly add to the visual appeal of a garden.

Birds feed on a wide variety of insects, depending mainly on the size of their bill. Occasionally greed will overcome a bird's better judgement, and it will attempt to eat an insect far too big for it to manage. Nevertheless, birds can swallow surprisingly large items with apparent ease.

Grasshoppers, when available, are an energy-rich source of food for a number of garden birds.

Leatherjackets are the larvae of the familiar crane-fly, or daddy-long-legs, and provide an important food source during the spring and summer.

Earthworms are hunted mainly by Blackbirds and Song Thrushes, which hop across the lawn, keeping a sharp eye open for worms coming to the surface.

Caterpillars are an important prey item for Blue and Great Tits, especially during the breeding season, when their plump bodies provide essential energy for the growing chicks.

Ponds will always attract insects, especially on warm evenings during the summer, when House Martins will often take advantage of the abundant supply of flying food! You will often see a House Martin swooping up quickly, snapping up an insect and then gliding back down.

GARDEN BIRDS
THROUGHOUT THE YEAR

NOT EVERY SPECIES OF BIRD IN YOUR GARDEN is found there all year round. Some, such as House Martin and Willow Warbler, are exclusively summer visitors to the British Isles. Others, like Redwing, Fieldfare and Brambling, are mainly winter visitors.

Although some resident species, such as Blue Tit and Robin, are present throughout the year, their numbers often increase in winter as a result of Continental immigrants, which move westwards to avoid the harsh winters of central Europe.

In recent years two species which were once primarily summer visitors, Blackcap and Chiffchaff, have begun to winter here in growing numbers. They are often found in gardens, where food may be more plentiful than elsewhere.

The table below indicates the peak times for seasonal visitors, as well as the status of all the species featured in this book.

KEY TO TABLE:

R = Resident (present throughout the year)
S = Summer Visitor
W = Winter Visitor

Months present indicated by numbers 1-12

Status indicated by type:
Bold = present in most gardens - e.g. **1 2 3**
Normal = present in some gardens - e.g. 1 2 3
Italic = rare or local visitor to a few gardens - e.g. *1 2 3*
( ) = occasionally seen during this month - (1 2 3)

SPECIES	Status	Months present											
Grey Heron	R	*1*	*2*	*3*	*4*	*5*	*6*	*7*	*8*	*9*	*10*	*11*	*12*
Mallard	R	*1*	*2*	*3*	*4*	*5*	*6*	*7*	*8*	*9*	*10*	*11*	*12*
Sparrowhawk	R	*1*	*2*	*3*	*4*	*5*	*6*	*7*	*8*	*9*	*10*	*11*	*12*
Buzzard	R	*1*	*2*	*3*	*4*	*5*	*6*	*7*	*8*	*9*	*10*	*11*	*12*
Kestrel	R	1	2	3	4	5	6	7	8	9	10	11	12
Red-legged Partridge	R	*1*	*2*	*3*	*4*	*5*	*6*	*7*	*8*	*9*	*10*	*11*	*12*
Pheasant	R	*1*	*2*	*3*	*4*	*5*	*6*	*7*	*8*	*9*	*10*	*11*	*12*
Moorhen	R	*1*	*2*	*3*	*4*	*5*	*6*	*7*	*8*	*9*	*10*	*11*	*12*
Black-headed Gull	W (R)	1	2	3	(*4*	*5*	*6*	*7*	*8*	*9*	*10*)	11	12
Common Gull	W	*1*	*2*	*3*	(*4*	*5*	*6*	*7*	*8*	*9*	*10*)	11	12
Stock Dove	R	*1*	*2*	*3*	*4*	*5*	*6*	*7*	*8*	*9*	*10*	*11*	*12*
Woodpigeon	R	**1**	**2**	**3**	**4**	**5**	**6**	**7**	**8**	**9**	**10**	**11**	**12**
Collared Dove	R	1	2	3	4	5	6	7	8	9	10	11	12
Feral Pigeon	R	**1**	**2**	**3**	**4**	**5**	**6**	**7**	**8**	**9**	**10**	**11**	**12**
Turtle Dove	S				(*4*)	*5*	*6*	*7*	*8*	(*9*)			
Cuckoo	S				(*4*)	*5*	*6*	*7*	(*8*)				
Barn Owl	R	*1*	*2*	*3*	*4*	*5*	*6*	*7*	*8*	*9*	*10*	*11*	*12*
Little Owl	R	*1*	*2*	*3*	*4*	*5*	*6*	*7*	*8*	*9*	*10*	*11*	*12*
Tawny Owl	R	*1*	*2*	*3*	*4*	*5*	*6*	*7*	*8*	*9*	*10*	*11*	*12*
Kingfisher	R	*1*	*2*	*3*	(*4*	*5*	*6*	*7*)	*8*	*9*	*10*	*11*	*12*
Green Woodpecker	R	*1*	*2*	*3*	*4*	*5*	*6*	*7*	*8*	*9*	*10*	*11*	*12*
Great Spotted Woodpecker	R	1	2	3	4	5	6	7	8	9	10	11	12
Lesser Spotted Woodpecker	R	*1*	*2*	*3*	*4*	*5*	*6*	*7*	*8*	*9*	*10*	*11*	*12*
Swift	S				(4)	**5**	**6**	**7**	**8**	(9)			

Bird	Status	1	2	3	4	5	6	7	8	9	10	11	12
Swallow	S			(3)	4	5	6	7	8	9	(10)		
House Martin	S				4	5	6	7	8	9			
Grey Wagtail	W (R)	1	2	3	4	(5	6	7	8	9)	10	11	12
Pied Wagtail	R	1	2	3	4	5	6	7	8	9	10	11	12
Waxwing	W	1	2	3	(4)						(10)	11	12
Wren	R	1	2	3	4	5	6	7	8	9	10	11	12
Dunnock	R	1	2	3	4	5	6	7	8	9	10	11	12
Robin	R	1	2	3	4	5	6	7	8	9	10	11	12
Fieldfare	W	1	2	3	(4)					(9)	10	11	12
Blackbird	R	1	2	3	4	5	6	7	8	9	10	11	12
Song Thrush	R	1	2	3	4	5	6	7	8	9	10	11	12
Redwing	W	1	2	3	(4)					(9)	10	11	12
Mistle Thrush	R	1	2	3	4	5	6	7	8	9	10	11	12
Garden Warbler	S				(4)	5	6	7	8	(9)			
Blackcap	S (W)	1	2	3	4	5	6	7	8	9	10	11	12
Chiffchaff	S (W)	1	2	3	4	5	6	7	8	9	10	11	12
Willow Warbler	S				4	5	6	7	8	9			
Goldcrest	R	1	2	3	4	5	6	7	8	9	10	11	12
Spotted Flycatcher	S				(4)	5	6	7	8	(9)			
Long-tailed Tit	R	1	2	3	(4	5	6	7	8	9)	10	11	12
Marsh Tit	R/W	1	2	3	(4	5	6	7	8	9)	10	11	12
Willow Tit	R/W	1	2	3	(4	5	6	7	8	9)	10	11	12
Coal Tit	R/W	1	2	3	(4	5	6	7	8	9)	10	11	12
Blue Tit	R	1	2	3	4	5	6	7	8	9	10	11	12
Great Tit	R	1	2	3	4	5	6	7	8	9	10	11	12
Nuthatch	R/W	1	2	3	(4	5	6	7	8	9)	10	11	12
Treecreeper	R/W	1	2	3	(4	5	6	7	8	9)	10	11	12
Jay	R	1	2	3	4	5	6	7	8	9	10	11	12
Magpie	R	1	2	3	4	5	6	7	8	9	10	11	12
Jackdaw	R	1	2	3	4	5	6	7	8	9	10	11	12
Rook	R	1	2	3	4	5	6	7	8	9	10	11	12
Carrion Crow	R	1	2	3	4	5	6	7	8	9	10	11	12
Starling	R	1	2	3	4	5	6	7	8	9	10	11	12
House Sparrow	R	1	2	3	4	5	6	7	8	9	10	11	12
Tree Sparrow	R/W	1	2	3	(4	5	6	7	8	9)	10	11	12
Chaffinch	R	1	2	3	4	5	6	7	8	9	10	11	12
Brambling	W	1	2	3	(4)					(9)	10	11	12
Bullfinch	R	1	2	3	4	5	6	7	8	9	10	11	12
Greenfinch	R	1	2	3	4	5	6	7	8	9	10	11	12
Goldfinch	R	1	2	3	4	5	6	7	8	9	10	11	12
Siskin	W/R	1	2	3	4	(5	6	7	8	9)	10	11	12
Linnet	W/R	1	2	3	4	(5	6	7	8	9)	10	11	12
Redpoll	W/R	1	2	3	4	(5	6	7	8	9)	10	11	12
Hawfinch	W/R	1	2	3	4	(5	6	7	8	9)	10	11	12
Yellowhammer	W/R	1	2	3	4	(5	6	7	8	9)	10	11	12
Reed Bunting	W/R	1	2	3	4	(5	6	7	8	9)	10	11	12

RARITIES AND ESCAPES

O NCE IN A WHILE, YOUR GARDEN MAY PLAY HOST TO AN UNUSUAL VISITOR. It may be an escaped cagebird, such as a Budgerigar or Canary, though these are unlikely to survive the British weather for very long. Hardier species may establish a thriving feral population, in the case of some, like the Ring-necked Parakeet, even becoming an agricultural pest.

Rarer visitors, too, may turn up from time to time. If you see an unusual bird in your garden, get in touch with the RSPB or the Bird Information Service (see page 174). Who knows, it may prove to be as rare as the Yellow-rumped Warbler which visited a Devon birdtable one winter. This was the first record of this North American species in Europe.

Ring-necked Parakeets, originally from Asia, were released by cagebird dealers back in the 1960s and 1970s. They have since established thriving populations in several parts of southeast England, and are frequent visitors to suburban gardens, where they are especially keen on fruit trees. Their emerald-green plumage, long tail, and distinctive screeching call are unmistakable.

Every autumn, rare wanderers from Siberia arrive in small numbers on our eastern coasts, thousands of miles away from their intended destination. Although most meet an early death, a few move inland, where they sometimes turn up in gardens. Mild weather may allow birds like this Yellow-browed Warbler to survive the winter, especially in the more sheltered parts of southwest England.

Hoopoes are rare but regular spring visitors to the British Isles, and may occasionally be seen on garden lawns, especially in southern England. They are about the size of a Mistle Thrush, but have a bright pinkish plumage, black-and-white wings and a noticeable crest. Beware confusion with the Jay, which can appear very pink when seen on the ground, but has a much shorter bill and no crest.

URBAN GARDENS

E ven if your garden is just a tiny patch of green in the centre of a city, it can still attract birds. Think of it as a welcome oasis amidst the concrete desert, providing shelter, food, and in many cases also a place to nest.

You can increase the number and variety of birds visiting your garden by making it as 'bird-friendly' as possible: providing birdtables, birdbaths and nestboxes, and trying to keep out unwelcome visitors such as cats and squirrels.

In early spring, birds such as House Sparrows are looking for bits and pieces with which to build their nests. In towns and cities these may be in short supply, but you can help by providing alternative materials such as wool or hair.

Hard winter weather often drives birds to seek refuge in urban gardens, where a guaranteed supply of food and water can make the difference between life and death. As well as larger numbers of the regular species, your garden may play host to more unusual visitors, such as Coal Tit, woodpeckers or this Nuthatch, seen here in characteristic pose on a nutbag. If harsh weather persists, these birds may remain in the garden for some time provided that food and water are constantly supplied.

During prolonged spells of freezing weather, waterbirds such as Moorhen, Water Rail or Snipe will occasionally visit gardens. The Snipe feeds mainly on worms, using its long bill to probe deep into the hard soil.

PARKS AND
OPEN SPACES

ONCE YOU HAVE BEGUN TO GET TO GRIPS WITH THE BIRDS IN YOUR GARDEN, you will probably want to venture further afield. A good place to start is your local park, where you can see a wider variety of species, and test your new-found identification skills.

Go during a busy summer weekend, and there won't be many birds around. But pay a visit early in the morning, before the children and dog-walkers are out and about, and you'll find a very different scene. This is when most birds are very active – feeding, singing and calling, bathing, preening and so on, and they are often relatively tame as well, so you can approach them more closely.

The Hawfinch is one of our shyest birds, and is best looked for at a known breeding or wintering site. Listen out for its distinctive call, an explosive 'zik'. Hawfinches are often seen singly or in small flocks near the tops of trees, their large size, 'bull-headed' appearance and prominent white wing bars being distinctive. The large bill is ideal for cracking cherry stones, in order to reach their kernels. They also eat the seeds of a number of trees, including the elm.

About the size of a sparrow, the Lesser Spotted Woodpecker is our smallest woodpecker, and by far the hardest to see. It is best looked for in winter, when it may join a flock of tits and other small birds in search of food. In spring, its soft drumming may be heard in wooded areas of large parks, though the species is highly vulnerable to human disturbance.

In spring, breeding birds will be singing before dawn. Bird sounds can be confusing for a beginner, so it is worth spending some time getting to know the different calls and songs, perhaps using Compact Discs or cassette tapes. Singing birds are often difficult to see because of dense foliage, but be patient, and they will eventually reveal themselves.

Larger parks and open spaces will support an even greater variety of birdlife, though you may have to search for some time before you see any of the shyer species.

If your local park has a pond or lake, this will attract ducks and other waterfowl, especially in autumn and winter. These birds are usually less prone to human disturbance than songbirds, and may allow very close approach, especially if you bring along some bread to feed them! The most likely species are Tufted Duck, Pochard, Mallard, Shoveler, Moorhen and Coot, and of course the ubiquitous Canada Goose.

The Little Owl is the most likely member of its family to be seen during the day, and is most active soon after dawn or before dusk. Look out for it sitting on a post or branch of a tree, or flying down to the ground to catch rodent or insect prey. Following numerous attempts to introduce the Little Owl into many parts of Britain, during the last century, it is now quite common on lowland farms and other similar habitats.

FEEDING BIRDS

TTRACTING BIRDS INTO YOUR GARDEN, HOWEVER BIG OR SMALL, is something that will bring you endless hours of enjoyment and pleasure. Roving flocks of birds pass through gardens with some regularity, but how do you keep them in your garden for that little bit longer? What can you do to ensure that the birds will keep coming back into your garden?

A guaranteed way of enticing birds into your garden over and over again is to feed them. As 'word goes around' that food is available in your garden, more and more birds will pay a visit and you should see a marked increase in activity. But what should you feed them and when? Are there particular foods that should be put out in winter and not in summer or vice versa?

This section will cover all those queries and quandaries, give you the best advice with regard to 'safe' foods and, it is hoped, some new ideas on keeping the birds in your garden fed and watered.

From birdtables to seed hoppers, birdbaths to 'tit bells' and scrap baskets, the following pages will look carefully at successful methods of feeding birds.

Robins are undoubtedly one of the most popular visitors to any garden. Highly adaptable, they will return to the garden for food whatever the time of year. If space is limited in your garden, place food into a terracotta bowl, which has the advantage of being frostproof. Time and time again birds will repay your patience by becoming more trusting and approachable.

A birdtable is one of the most traditional and familiar methods of attracting birds into your garden. Birdtables come in all sorts of shapes and sizes – with or without sides, with or without perches, sometimes with a roof (often thatched), and some even have a nestbox included! As you can see from this familiar scene, birdtables can attract many different species – here we have Chaffinch, Great Tit and House Sparrow – and they can all gain plenty of nourishment from foodstuffs left out for them. Clearly they will feed on different items, but do try to provide for them all.

In winter months you can gain so much pleasure from gazing out of the kitchen window (never mind the washing-up!) watching birds squabble and feed on a nutbag. Here we see a Siskin encountering a Great Tit and a Blue Tit, all of them vying for the choicest peanuts.

Water is as vital to a bird's survival, particularly in winter, as a good store of food. A birdbath (use anything from an old dustbin lid to a purpose-built birdbath) will provide birds with a place to bathe and keep their feathers in good working order, as well as providing them with an essential source of drinking water. This Song Thrush is making full use of the bath, which in winter is kept ice-free by the 'night light' placed underneath.

Some species are not too fussy about their diet, and will eat more or less whatever you give them. These Starlings are feeding on kitchen scraps, hung up in a basket to avoid attracting rats or mice.

Seeds are the staple diet of finches such as Greenfinch and Goldfinch. Their different-shaped bills are specially adapted to eat particular seeds, so make sure that you provide a variety of shapes and sizes – and in reasonable quantity if you wish the birds to return!

One of the best ways to observe birds is by using a window feeder. Birds like this Robin and Blue Tit will soon get used to it, allowing you to watch them in dramatic close-up from the comfort of your living-room!

A hollow stick pierced with small holes and filled with suet, available from many good bird-food suppliers, may attract scarcer species such as Great Spotted Woodpecker and Nuthatch to your garden, particularly in the winter months and especially if you live not too far from a wood, even a small one. This kind of feeder is also, of course, attractive to tits.

A tit bell crammed with fat or suet is a good way to provide a high-energy food for these tiny birds, which may need to eat as much as a quarter or even more of their body weight every day in order to survive cold weather. In very harsh winters many tits would perish without the provision of such food in gardens.

Mealworms, available from pet stores and fishing supply stores, are an ideal high-energy winter food for many species of bird, including tits, thrushes and this hungry Blackbird.

NEST SITES

ONE OF THE BEST WAYS OF ATTRACTING BIRDS TO YOUR GARDEN is by providing sites where they can raise their young. Trees, shrubs, holes in the wall, even old drainpipes, can all be suitable places to build a nest. Even a small garden can support as many as half-a-dozen nests, while a larger garden, with a variety of habitats, is a valuable refuge for breeding birds.

Keen gardeners can encourage birds to nest by planting a variety of suitable shrubs and trees. Native plants such as Elder and Hawthorn are usually best, although fast-growing evergreens such as cypresses can provide cover, too.

Birds do not only build their nests in natural sites, and will take readily to man-made nestboxes. These come in all shapes and sizes, each designed to attract a particular species while preventing others from using the box.

It is usually best to put up a nestbox during the winter, so that the birds get used to it before the breeding season begins. The best site is on a tree, wall or garden fence, between two and five metres above the ground. Nestboxes should generally face between northeast and southeast, thus avoiding the midday sun and the wettest winds. Try to place the box out of the reach of potential predators and inquisitive humans!

The RSPB provides fact sheets containing information on the several different kinds of nestboxes, with details on how to build your own. It also sells ready-made boxes by mail order, if DIY isn't your speciality! *(See page 174)*

The most widely used nestboxes have a small opening, around 25-32 mm in diameter, depending on which particular species you want to attract. Great Tits prefer a hole around 28 mm across. They produce one or two broods, with between five and eleven eggs, which the adults incubate for around a fortnight. The young fledge three weeks later. You will be able to watch them making frequent sorties to provide moth caterpillars and other insects for their hungry, new brood.

An open-fronted nestbox is suitable for species such as Robin, Wren and Spotted Flycatcher. Wrens usually lay two clutches of between five and eight eggs, although the male often has to build several nests before one is selected by his choosy mate! During cold snaps in winter, Wrens sometimes roost in nestboxes, with as many as several dozen individuals huddled together in a single box.

This specialised nestbox, made to resemble a natural crevice in a tree trunk, is designed to attract the Treecreeper. In woodlands, a typical natural site would be behind a loose piece of bark or in ivy. This tiny, mouse-like bird is sometimes found in rural gardens, where it can be seen climbing up tree-trunks in search of its insect food.

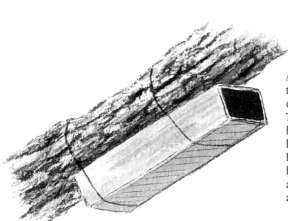

A decline in the number of fallen and rotten trees has drastically reduced the availability of nest sites for larger birds such as the Tawny Owl. You can help redress the balance by providing specially designed owl boxes, which are fixed to the underside of branches or placed in a tree fork. These boxes are best placed in an area where owls are known to be resident – perhaps where an existing site has been destroyed.

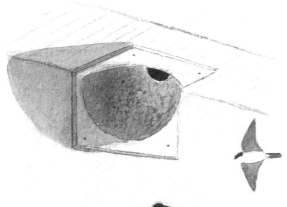

House Martins are one of our most familiar summer visitors, arriving in late April and departing for their African winter quarters in September and October. As their name suggests, they usually build their nests under the eaves of houses, using pellets of wet mud. Nowadays you can encourage House Martins to breed by putting up artificial nests, though your neighbours may complain, as the young are very noisy!

Kestrels are our commonest and most familiar bird of prey, nesting in a variety of habitats including tree holes and the ledges of buildings. They require a large, solid nestbox, out of reach of human interference. The open-fronted design allows you to observe the young as they grow, and you can also see the type of food the parents bring in for their brood.

The Spotted Flycatcher is one of our daintiest and most interesting summer visitors, living up to its name by making acrobatic chases after its insect prey. It nests in a variety of sites, but is often attracted to hanging baskets, as well as ledges and specially designed nestboxes. It is streaked rather than spotted, but the streaks are faint and are restricted to the crown and breast. Juveniles, however, look much more spotted, including over the entire head and mantle.

GARDEN HAZARDS

ARDENS MAY BE A WELCOME HAVEN FOR THEIR OWNERS, but for birds there are few more hazardous places. The greatest villain of the piece is the domestic cat, responsible for millions of bird deaths every year. Unfortunately, the majority of cats' victims are adult birds, whose death often results in nesting failure.

Some of the major predators are other birds, attracted to gardens by the large concentration of smaller birds there. To a bird of prey like the Sparrowhawk, a birdtable looks like a free lunch.

You can take steps to reduce the carnage in your own garden by making sure that food is placed on sturdy birdtables, out of the reach of most predators.

Perhaps the most notorious garden predator of all is the Magpie. However, Magpies are not quite the villains they are sometimes painted as. Unlike cats, Magpies generally prey on eggs and chicks, so, even if a brood is destroyed, the adult birds have the chance to raise a second one. Also, the vast majority of songbirds die during their first year of life anyway, so the Magpie's catch does not have much effect on the population as a whole. Finally, it should be remembered that Magpies, in contrast to cats, are at least part of the natural food-chain!

Watch a cat stalking a bird, and you can see its wild ancestry at a glance. Unfortunately, cats have a very damaging effect on the populations of many garden birds, especially during the breeding season. One way to thwart the cat's murderous attempts is to tie a bell around its neck. This generally warns the birds of the cat's approach, and should significantly reduce the death toll.

A tree-hole full of Blue Tit chicks is a tasty meal for a Great Spotted Woodpecker. The chicks' only hope of survival is if the entrance to the hole is too narrow for the hungry woodpecker to enter.

One of the most spectacular and memorable sights of garden birdwatching is a Sparrowhawk in pursuit of its prey. The hawk's short, blunt wings and long tail enable it to approach fast and unseen, at a low angle. This unsuspecting Chaffinch may never have known what hit it.

There is nothing a squirrel likes quite so much as a meal of eggs or young birds. The best way to make your nestbox squirrel-proof is by fixing a sturdy metal plate over the area around the hole. This should foil all but the most determined intruder.

SECTION TWO:

IDENTIFYING THE BIRDS

GREY HERON
(90-100CM, 36-40IN)

THE GREY HERON IS AN UNMISTAKABLE BIRD, the largest common land bird that can be seen in Britain and also the most widespread heron in Europe. It can be found along almost all waterways, from lakes, rivers, marshes and estuaries to the simple garden pond, where, with its fish diet, it is not the most welcome of visitors!

Grey Herons are tall, slender, elegant birds with a heavy, dagger-like bill, long legs and grey, black and white plumage. When fishing knee deep in water, an adult Grey Heron presents an impressive sight. The long neck is fully extended and the stealth of such a big bird is a marvel.

In flight, the size of Grey Herons really becomes apparent. The overall appearance is of a huge greyish bird flying very slowly on bowed wings with a series of deep wingbeats, its neck held hunched and long legs trailing. The underwing will appear blackish and the black flank line will be quite obvious, even at considerable distance.

The adult's head is white, with two broad black stripes running over the back of the crown and extending, in the breeding season, into slim black head plumes. The neck is white, washed pale grey on the sides, with black flecks running from the throat to the belly. Breeding birds have shaggy plumes on the lower breast. The flanks are black, with white on the belly extending to the vent, with a dark undertail. The upperparts are blue-grey, with a black 'shoulder' and pale silvery plumes over the wing. The bill is orange-yellow. The eye and legs are yellowish.

Juvenile Grey Herons always appear slightly stockier than adults and are always darker, usually grey but sometimes brown. The head lacks the white and black badger stripes of the adult, the forehead, crown and nape all being dark grey, with only a white throat for contrast. The upperparts and underparts are generally dull grey, lacking plumes and shoulder patch. The bill is brownish-grey with a yellowish-grey lower mandible. The eye is dull yellow and the legs are olive.

▲ Grey Herons will invariably call in flight, a loud hard 'frank' or 'krank'.

▶ Grey Herons are communal nesters often favouring woodland trees as a breeding site, but they also breed in reedbeds and, occasionally, on cliffs.

The nest-building and display rituals start in late January and, when presented with a suitable stick or twig, much bill-snapping and calling breaks out.

◀ This juvenile is captured in classic pose, waiting to seize the moment and pounce on its prey.

◀ A masterful-looking bird, the Grey Heron looks best when standing serenely on a boulder, mid-river, surveying its territory, plumes flapping gently.

▲ Grey Herons, when sleeping, can be surprisingly easy to overlook, particularly if asleep on a reedbed fringe. Seemingly oblivious to disturbance, they can remain hunched up for long periods of time.

MALLARD
(55-62CM, 22-25IN)

THE MALLARD IS A VERY FAMILIAR AND VERY ABUNDANT DUCK, that is found throughout Europe. It can be found in any watery location, from secluded streams to the local boating lake. Mallards are often very tame and will show no fear whatsoever in approaching people.

A hefty large-headed and long-billed dabbling duck, the Mallard is prone to interbreed with 'farmyard' ducks, and so produces all manner of peculiar-looking offspring. Pure Mallards are, however, easily identified, particularly the male, and even the brown female should not really present a problem: size alone rules out other dabblers. The female is responsible for the familiar 'quack quack' call, whereas the male gives a low nasal whistle.

The most likely species to be confused with the female Mallard is the female Gadwall. Gadwalls are smaller, slimmer and leaner looking than the portly Mallard and show a distinctive white speculum, whatever the view. The bill of the Gadwall is also a prime identification feature – dark grey on top with yellowy-orange sides.

The female Mallard is a decidedly nondescript affair but can be quite variable. The head is usually brown, with a darker crown and a dark line through the eye. The upperparts and underparts show a variegated brown and black pattern. The bill is dark brown, with orangey edges.

The male Mallard is unmistakable with a bottle-green head separated from the rusty-brown breast by a distinctive white neck ring. The bill is a bright banana-yellow colour. The eye is black and the feet are orange in both sexes.

► When seen flying, the male Mallard (right) shows a bulbous green head on the end of the long rusty neck. The upperparts are greyish, except for the black rump 'wedge' and white tail. The upperwing is grey, except for a purple speculum along the rear wing. Two white wing bars are apparent. From above, a female (left) will look brown and black except for the dark tail, white wing bars and dark speculum.

▼ A familiar view of Mallards is to see them upending in search of food – the male is recognisable by his orange legs and, particularly, his black and white rear end. The female exhibits a white edge to the tail and blackish spots on the undertail.

▼ In summer months a Mallard can change appearance quite markedly. A male in eclipse plumage shows a pale brown head, with black crown and eyestripe. The upperparts change to blackish-grey, except for the paler flight feathers, and the rusty-brown breast becomes scalloped black. The bill becomes dull greyish-yellow. A female in June to September will hardly change: the crown gets darker and the upperparts are more uniform in tone.

SPARROWHAWK
(28-38CM, 11-15IN)

The Sparrowhawk is a smallish, agile bird of prey which is widespread across Northern Europe. Typically a bird of open countryside, woodland and hedgerows, the Sparrowhawk is now an increasingly familiar visitor to both rural and urban gardens.

Male and female Sparrowhawks differ markedly in plumage markings and size but they both share broad, blunt, shortish wings, a small-headed look and a long square-ended tail. They both have a dark-tipped hooked beak with a yellow base, yellow-orange eyes and yellow legs and feet with dark claws. Their appearance and dashing flight readily distinguish them from other birds.

A male Sparrowhawk is clearly smaller than the female. The upperparts are wholly blue-grey except for white lores and supercilium, russet cheeks, dark bars on the tail and dark wing tips. The underparts are finely barred orangey-red from the throat to the belly. The vent and undertail-coverts are white, and the undertail itself is greyish with darker barring. The white tip to the tail is also visible.

Female Sparrowhawks are the larger of the two sexes. The upperparts are wholly dark grey-brown except for a white supercilium and blackish-looking bars on the tail. The underparts show a white base colour and fine grey horizontal barring from the upper breast to the belly. The undertail-coverts are white. The tail is whitish with darker bars. Females tend to have a more striking yellow eye than the male Sparrowhawk.

The bluey upperparts and finely barred orange underparts are easily seen when a male Sparrowhawk flies. The underwing shows greyish barring on the secondaries and primaries. In flight, the wing shape, size of the head and tail are to be noted.

The dark brown-grey upperparts and strong tail bars of the female are very obvious in flight. The fine barring of the belly and forewings contrasts with the coarse markings on the remainder of the underwing and undertail. Sparrowhawks fly with quick bursts of rapid wingbeats interspersed with short glides. When soaring, they look flat-winged and the tail is only occasionally fanned.

Sparrowhawks are particularly adept at dashing through woodlands, along hedges and through gardens at great speed, and if successful will often take the prey to a plucking post. On the post, the Sparrowhawk will tear feathers from the corpse and take flesh back to a mate or to youngsters in a nest. If a post is not available, a Sparrowhawk will often perform its grisly act at the scene of the capture.

Juvenile Sparrowhawks can be recognised and separated from adults by browner upperparts, often with fresh rufous edges, and brownish-buff barring on the underparts, although juvenile males tend to show some rufous on the belly and breast. The eyes are greener than those of adults.

BUZZARD
(50-56CM, 20-22IN)

THE BUZZARD IS A CHARACTERISTIC, BROAD-WINGED BIRD OF PREY which is fairly widespread across much of western and northern Europe. It breeds across much of western England, Wales and most of Scotland and Ireland. In Continental Europe, Buzzards are commonplace, except in northern Scandinavia, where they are largely absent. The Buzzards in northeastern Europe move south and westwards during the autumn and winter, sometimes in quite large numbers.

The Buzzards' favourite habitat is mixed woodland adjacent to farmland, giving them the ideal opportunity to nest and feed in close proximity, but they can also be seen in moorland and upland areas. In more rural areas of Britain they can frequently be seen drifting over gardens, and even above town centres!

Buzzards come in a variety of different plumages, and have a seemingly effortless flight manner, spiralling into the sky. They can often be detected by their cat-like 'me-uw'.

This Buzzard shows one extreme of their plumage. It is particularly pale on the underparts, with slight brown flecking on the breast sides. The head is noticeably paler than on the typical bird (shown below), with prominent white supercilia and throat contrasting with the brownish ear-coverts.

This familiar view of the Buzzard on an old gatepost shows a typically plumaged individual. Most of the plumage is dark brown, except for paler ear-coverts and a whitish throat and lower-breast patch. The undertail is also marginally paler, usually dark greyish, with finer darker barring. The stout, hooked bill is black with a yellow cere. The eye is dark brown, and the legs and feet are bright yellow, with black claws.

▲ When seen flying towards you, particularly when gliding, the wing profile of the Buzzard is highly distinctive. The wings are always held slightly *above* body level, with the wing tips slightly upturned.

◀ The broad wings, large head and shortish tail are easily seen when a Buzzard soars overhead. This dark bird shows a number of characteristic plumage features, including the brownish head, breast and belly contrasting with the pale undertail. Notice the distinctive three-toned wings.

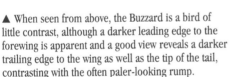

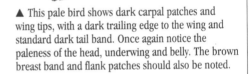

▲ This pale bird shows dark carpal patches and wing tips, with a dark trailing edge to the wing and standard dark tail band. Once again notice the paleness of the head, underwing and belly. The brown breast band and flank patches should also be noted.

▲ When seen from above, the Buzzard is a bird of little contrast, although a darker leading edge to the forewing is apparent and a good view reveals a darker trailing edge to the wing as well as the tip of the tail, contrasting with the often paler-looking rump.

▶ Buzzards are good opportunist feeders, killing for themselves or taking full advantage of a death from natural causes. Sadly, this means that this glorious species is susceptible to much persecution from certain people. Poisoned bait is a favourite method of bringing a slow painful death to a splendid bird.

41

KESTREL
(33-36CM, 13-14IN)

$\mathbf{T}$HE KESTREL IS THE COMMONEST BIRD OF PREY that can be seen in Britain, and is widespread throughout western and northern Europe. Kestrels are found in any number of different habitats, from the largest city to remote hillsides.

Kestrels are perhaps most commonly seen on roadsides, hovering in search of food or perched on roadside wires, telegraph poles and fences.

These distinctive falcons are easily told from other birds of prey by their size, longish tail and noticeably pointed wings. Males and females appear, at long range, to be very similar, but a good view will show the numerous differences between them. However, both share a small hooked bill with a grey-black tip and yellow base, large black eye with yellow eyering and yellow feet with black claws.

▶ Females are easily recognised by their larger size and shape alone, but the rufous and dark barred colouring makes them unmistakable. The head shows a brown crown and nape with fine dark streaks. The cheeks and throat are white with a prominent black moustache. The rufous mantle has dark bars, extending to the wing-coverts. The barred rump is slightly greyer brown. The tail shows six to seven prominent dark bars and a white tip.

◀ The male's head is blue-grey, with slightly paler cheeks, dark moustache and buff throat patch. The mantle and wing-coverts are chestnut with black spotting. The rump and tail are pale blue-grey, except for the broad black wing bar. Note the pinkish-white underparts showing distinctive black 'teardrops' from upper breast to belly. The undertail-covert area is whitish. The undertail is greyish with a black tip.

In flight the differences between the sexes are clear. The grey of the male's head and tail contrasts strongly with the chestnut and black of the wings, while the overall rufous tones of the female on the upperparts are easily seen. Check the tail pattern, too – look for all those bars.

This is the classic view of the Kestrel – head down, tail forward, flapping when necessary, hovering into the wind. When not in this position, Kestrels soar on wings that are held forward and a tail which is spread.

Kestrels have a whip-like, deep wingbeat in 'normal' flight, interspersed with glides.

Juvenile Kestrels basically resemble the female, except for a brighter, redder-looking rump and bolder streaks on the underparts. The bare parts are much as those of the male and female, although the legs and feet may appear more orange in colour.

RED-LEGGED PARTRIDGE
(33-36CM, 13-14IN)

THE RED-LEGGED PARTRIDGE IS A COLOURFUL GAME BIRD which is becoming a familiar sight in southern England. Its range outside Britain is restricted to southwestern Europe. Red-legged Partridges can be found in a wide variety of habitats -- farmland, heathland, shingle beaches and dunes, so, if you have a garden in any of these habitat areas, do not be surprised to see one of these birds wander across the lawn!

Red-legged Partridge is a rounder-looking bird than the Grey Partridge, and often the Red-legged appears more upright than the Grey. 'Redlegs' show a more striking head pattern, a plainer back and more complex underpart markings than Grey Partridge.

The bill is short and cherry-red in colour. The eye is dark brown with a red orbital ring, while the legs are, of course, red.

Red-legged Partridge shows a grey-brown crown and nape, contrasting markedly with the bold white supercilia and throat patch. The cheeks are brownish, while the lores and border of the throat are black. The black becomes spotted on the ear-coverts and upper breast, so forming an obvious 'necklace'. The upperparts, with the exception of dark wing tips and orangey tail sides, are dull olive-brown.

The upper breast of the Red-legged Partridge is warm brown, merging into an ashy-grey lower breast, fading to a rich orange on the belly and undertail. The flanks are a mix of black, white, red and blue bars.

The whirring wings are always a giveaway! Red-legged Partridges fly quickly and low on decidedly bowed-looking wings. As they fly away from you, the orange on the tail is particularly striking. Red-legged Partridges do have a habit of running more than flying, and, as they run, the loud rasping 'chuck, chuck, chukak' call will usually be heard.

When seen from behind, the Red-legged Partridge remains utterly distinctive. The back of the head is strongly marked with white and black, while the flank bars are easy to see. The short wings reveal the buffy rump and orangey tail.

The juvenile Red-legged Partridge resembles a pale, faded version of the adult. The face pattern is very faint, the supercilia and throat are buffy-white, the 'necklace' is replaced by sparse grey mottling, while the rest of the head, mantle, rump and central tail are dull grey-brown. The wings are blotched with black and white. The outer tail is chestnut.

This juvenile shows the buff-brown upper breast, the pale blue lower breast and pink belly. The flanks are barred whitish and brown. The bare parts show a greyish bill, black eye and dull fleshy-pink legs.

PHEASANT
(52-90CM, 20-35IN)

THE PHEASANT IS AN UNMISTAKABLE, LONG-TAILED GAMEBIRD and is common throughout Britain, western and northern Europe. Pheasants are found in woodland, farmland, marshes and even reedbeds. They are often seen, particularly in the early morning or late evening, wandering serenely across gardens in more rural areas.

Pheasants were introduced into some parts of Europe, from Asia, as long as 2000 years ago, and have been present in northern Europe for over 200 years. Many thousands are released, for shooting purposes, on estates throughout Britain and the interbreeding can produce a wide variety of colour forms. Particularly attractive is the melanistic form, not uncommon in some areas, in which the male has the entire plumage apart from wings and tail strikingly blackish, with a strong green gloss to the upperparts and a blue-purple gloss below. Melanistic females also appear a much darker rufous-brown.

Female Pheasants are highly variable creatures, although basic plumage patterns are similar whatever the colour. They are always smaller and shorter-tailed than the male.

Male Pheasants are also variable in appearance, but the most commonly encountered are the birds with distinctive white neck collars shown here.

When they emerge from cover, Pheasants make a great deal of noise – crashing and clattering from the undergrowth, whilst calling their distinctive rapid fire 'kutuk, kutuk, kutuk' as they go. Their flight pattern is short and low to the ground, with rapid flaps and long glides.

This female shows a brownish crown and nape with fine black bars. The face is buffy-brown with a dark mask which contrasts with the pale, semi-neck collar. The mantle, rump and wings are variably marked blackish on a fawn base colour, and the uppertail shows several black bars, widely spaced. The underparts are pale buff, with dark marks on the flanks and undertail.

This male bird shows a deep bottle-green head and 'ear tufts' with an obvious patch of red facial skin. It lacks the white neck collar. The mantle is a rich orange-red colour, variably blotched and spotted with cream, black and white. The rump is greyish-brown and the long spiky tail is warm brown, with copper-toned outer feathers. The central feathers show thin, broadly spaced bars to the tip.

47

MOORHEN
(31-35CM, 12-14IN)

THE MOORHEN IS ONE OF THE MOST DISTINCTIVE WATERBIRDS to be seen in Europe, commonly encountered on ornamental lakes, village ponds, rivers and garden streams throughout northern Europe.

Both sexes share the same plumage and soft-part characters. The head and underparts are a dark greyish-blue, in some lights appearing black or a glossy deep navy-blue. The upperparts are a deep brown, with the short wings tipped black. Along the flank is an obvious white line, which appears very bold against the dark wing and body feathers. The 'rear-end' of the Moorhen is blackish except for a gleaming white undertail. The frontal shield and bill (except for a yellow tip) are cherry-red, as is the eye. The juvenile Moorhen has a pinkish bill, often with a greenish frontal shield.

As these are highly territorial birds, Moorhen fights are commonplace, whatever the season. Bodies will tilt backward into the water, wings will flop and feet will become jousting weapons . . . and the noise! Explosive 'gurgling' 'kurr-ucks', 'kek's and 'kikik's litter the air as birds vent their anger. Conflicts can last many minutes and injuries, even fatalities, are not unknown.

On land, where they look a little ungainly and timid, the yellowy-green legs and large feet are obvious. Just above the knee a small red 'thigh' patch is often visible.

48

▶ In flight, Moorhens present a rather awkward spectacle. Their short wings and bulky body mean that flight is difficult and they appear very weak on the wing. Their long legs trail well beyond the tail tip.

◀ Adult Moorhens are good parents, taking great care of their youngsters. Throughout the breeding season, at the edge of ponds, rivers and reedbeds, adult birds can be seen gently passing food to their bald-headed fluffy black chicks.

▼ Young Moorhens show none of the adults' colour. The head is dark brown with a greyish-brown wash on the face. The chin and throat are white. The upperparts are brown.

▼ Looking much more at home on the water, the Moorhen moves with jerky actions and a constantly flicking tail, the white and black of which are visible from considerable range. Often they dwell to pick at the water surface, feeding on weeds, insects or seeds.

BLACK-HEADED GULL
(35-38CM, 13-14½IN)

T HE BLACK-HEADED GULL IS AN EASILY RECOGNISED SMALL GULL and is common throughout Europe. Although still thought of by some as just a seaside nester, Black-headed Gulls breed in almost any wetland habitat from inland sites such as gravel pits to coastal marshes. In the winter, Black-headed Gulls can be encountered anywhere, even stealing food from birdtables!

Black-headed Gulls are slim-looking birds with a domed head, slender bill and clearly pointed wings. There are distinct summer and winter plumages for the adults and immatures. Black-headed Gulls fly with a buoyant jaunty action with fast wingbeats. They are also adept at gliding, soaring and hawking (at great height) for insects.

A first-summer Black-headed Gull resembles a 'mix' of adult and immature plumage, moulting out of its first-winter plumage between February and April. The head shows a brownish hood, flecked with varying amounts of white. The wings show some traces of its juvenile plumage and the tail retains a dark terminal band. The bill has a dark tip and orangey base, while the legs are also orange.

Adult summer Black-headed Gulls are very easy to identify, and the distinctive plumage is generally attained in February or March. The head shows a chocolate-brown hood, with a partial thin white eyering. The hindneck is white. The mantle and most of the wing are silvery-grey, looking very pale in some lights. The flight feathers are white with bold black tips. The rump and slightly rounded tail are white, as are the entire underparts. The bill, legs and feet are a rather dark reddish, while the eye is black.

◄ When a first-winter 'Blackhead' is seen in flight, the brown on the wings is seen easily, contrasting with the grey mantle and inner wing. Note the broad blackish trailing edge on the upperwing. The black tail band is also clear.

▲ Both plumages share a very distinctive upper- and underwing pattern. Notice on the upperwing the strong white leading edge, contrasting with the black trailing edge to the outer wing feathers – the primary feathers. The underwing is silvery-grey except for a large dusky panel towards the end of the wing, while the white leading edge and black trailing edge are still obvious especially on adults.

◄ First-winter Black-headed Gulls resemble adult winter birds except for more mottling on the head, and the wings show a large amount of brown on the coverts, secondaries and tertials. The tail shows a broad black terminal band. The bill shows a muddy-orange base and a black tip, and the legs are also muddy-orange in colour. The eye is black.

▶ An adult winter Black-headed Gull shows identical plumage to the summer bird except for the head pattern. In winter plumage, Black-headed Gulls show a white head, with a dark spot behind the eye which appears very smudgy and two dark bands that run up and across the crown. The bill colour becomes darker and duller on the base and the tip can look blackish. The leg colour becomes dull orangey-red.

THE COMMON GULL IS A STRIKING, MEDIUM-SIZED GULL that is widely found throughout northern Europe, although, in Britain, Common Gulls breed only in Scotland, Ireland and northern England. They are, however, far more numerous throughout Britain in the winter months, and are quite distinctive with their dirty-yellow bill, legs and feet and beady black eye.

They can be encountered in almost any location, urban or rural, coastal or inland. Common Gulls are frequently seen in the company of Black-headed Gulls, particularly in the winter months when they search for worms and insects on newly ploughed fields. They are, of course, more than happy when they search out fish and other aquatic life in coastal habitats. A medium-sized gull, the Common Gull is clearly larger than the very common Black-headed Gull, but is noticeably smaller than the two species of black-backed gulls and the archetypal 'seaside' gull, the Herring Gull. Adult Common Gulls have the unique combination of wholly yellow bill and legs.

First-winter Common Gulls are relatively easy to identify. The head shows heavy streaks on the crown and cheeks. The mantle is dark grey and the wings show brown coverts and tertials, while the flight feathers are black. The rump and tail are white, flecked black, with a broad black terminal tail bar. The underparts from lower breast to undertail are white.

This winter adult shows a white head with light brownish-grey flecking and streaks, which often extend to the upper breast. The mantle and wings are deep pearly grey, except for the double white crescents and the black and white primary feathers. The rump and tail are white. The underparts are wholly white.

► In flight, the young Common Gull is full of contrast. Note the streaked head, the dark back, the black, brown and grey wings, the white rump and the big tail band.

► In flight, the dark wings of the winter adult (right) are obvious, with striking black wing tips with bold white marks. Look also for the broad white trailing edge along the wing. A second-winter bird (above) differs from the adult winter in having more brown on the outer wing, heavy greyish blotching on the head and duller, more olivey-coloured bare parts.

▼ In the summer, the head is gleaming white and the bill and feet are bright yellow. The eye shows a blood-red orbital ring.

THE STOCK DOVE IS A VERY NEAT, COMPACT BIRD which is found commonly across most of Europe, with the exception of northern Scotland and Scandinavia. It tends to favour woods and open farmland, but it can still be found in more urban parks and gardens.

Stock Doves tend to be a little more solitary than other doves and pigeons, usually seen in pairs, but they do sometimes flock.

The Stock Dove is smaller than the Woodpigeon, with a small, squarish head, plump body and beautifully subtle plumage. It never shows the more obvious white neck patch and white wing patch of the adult Woodpigeon, and, unlike the latter, has two short dark bars across the inner wing. The two species appear quite different when seen together.

Stock Doves have a dark grey back, rump and wings, except for black on the coverts and primary feathers. The tail shows a broad black tip.

Stock Doves show a dark silvery-grey head with an emerald-green neck patch and salmon-pink breast patch. The underparts are quite variable, generally looking grey, but sometimes very blue-grey. The bill is pale yellow, with a whitish 'knob' and reddish base.

◄ When flying, the soft tones of the upper- and underparts are offset by the broad black trailing edge on the upper wing and the double spots on the coverts. The black tail band is also very obvious. White edges on the outermost tail feathers may be seen. The underwing will appear pale grey.

▶ Stock Doves hold their wings straighter than Woodpigeons and have a quick, direct flight, with occasional flicks.

▲ Stock Doves like nesting in natural tree holes, or sometimes in and around derelict buildings. The male will try to woo a female with his monotonous 'coo-oh, coo-oh' call.

55

WOODPIGEON
(39-45CM, 15-18IN)

THE WOODPIGEON IS A FAMILIAR, PLUMP BIRD which is very common across the whole of Europe. It can be found in almost any sort of habitat, but favours woodlands, gardens and parkland, whether in an urban or a rural situation.

Woodpigeons are very prolific birds, producing youngsters from March through to late November if the weather is suitable, and in the autumn months huge flocks containing many hundreds can be seen in agricultural areas.

Woodpigeons have a small head, a very full-chested look, broad wings and a longish tail. The bill has a dull reddish base, yellow tip and white cere. The eye is pale yellow and the feet and legs are pinkish. Their plumage pattern is distinctive, both in flight and on the ground.

◄ The juvenile Woodpigeon lacks the greenish gloss on the hindneck as well as the white neck patch. The breast is duller, showing a more buff-pink colour. The eye is a darker orange-yellow than the adult's, although other bare parts remain the same.

▶ An adult Woodpigeon's head is pale blue-grey, with a greenish gloss on the hindneck, a white neck patch and blue-grey nape and rump. The closed wing is darker grey, with white on the forewing and black flight feathers. The tail is grey with a broad black terminal band. The breast is purplish-pink, fading to white on the belly area.

Large flocks of Woodpigeons are a common sight. They take off with an almighty clatter as they leave fields or trees. They are capable of quick flight, with very deep wingbeats.

When seen from above, the white neck collar and large white wing patches are immediately apparent. From below, the most obvious features are the pale grey underwing, the pink breast, the off-white belly and the tail band.

A common sight in rural areas of Britain during the early autumn is that of flocks of Woodpigeons out and about feeding on fields of cut corn. Farmers will go to great lengths to deter the pigeons, but these seldom have any real effect.

COLLARED DOVE
(29-32CM, 11-13IN)

THE COLLARED DOVE IS A DISTINCTIVE, SLIM, LONG-TAILED BIRD which is widespread, and easily seen, across Europe. It can be seen in gardens in cities, towns and villages, as well as more open aspects.

The spread of Collared Doves is truly remarkable. Originally found in Asia, the 'invasion' into Europe really took off in the mid 1900s. By the early 1950s the first birds were seen in Britain, and by 1955 a small breeding population was becoming established in East Anglia. Now, some 40 years further on, the Collared Dove is found throughout Britain, in large numbers.

Collared Doves are fairly unmistakable birds, although a poor view could cause confusion with the Turtle Dove. Collared Doves are broad-winged and long-tailed, with generally pale plumage. The distinctive call of these doves, 'coo-coo-kut', with the middle syllable higher-pitched and stressed and the final one short and clipped, has become one of the most familiar natural sounds of suburbia. The call is sometimes mistaken for that of the Cuckoo, but there is really little similarity between the two. Collared Doves also give a drawn-out nasal 'kwurr' in flight.

Both adults and juveniles share a blackish bill, black eye with a thin white orbital ring and reddish legs and feet. Similar in many ways, this juvenile (right) lacks the familiar collar of the adult, and has greyer, more scaly-looking plumage.

The head and breast are a delicate shade of buffy-pink. The neck shows the distinctive black and white collar. The mantle is pale sandy-brown, the rump grey, fading to buffy uppertail feathers. The closed tail looks greyish, with buffy and white outer feathers. The wings are sandy-brown, with grey on the forewing and black wing tips. The belly is pale ashy-grey, with a hint of pink, becoming darker on the undertail-coverts and vent. Some Collared Doves have extremely pale plumage, looking very 'washed-out'.

◄ Collared Doves spend a lot of time perched on wires or telegraph poles. Males will 'coo' to attract a mate, while both sexes will make themselves comfortable, tucking heads in, seemingly just passing the time of day without a care.

▲ The pale head and black collar are clearly visible when Collared Doves are seen in flight. However, a view of the upperside of the bird will reveal some more striking features. The black wing tips contrast strongly with the grey and brown inner wings. The grey rump is very obvious, but even more striking is the tail pattern – white tips to all but the central feathers, with blackish grey at the base.

◄ Collared Doves spend some time drifting on their broad wings. When seen from below, they will look very pale: the breast will still look pink, but this contrasts with the greyish-white underwings and the broad black and white pattern on the undertail.

FERAL PIGEON
(31-35CM, 12-14IN)

The Feral Pigeon is a very familiar, round-looking dove which is extremely common across the whole of the continent. These birds can be seen absolutely anywhere, being incredibly easy to see in towns and cities, often in very large numbers. They are renowned for their extreme tameness, perfectly happy taking food from the hand. They are also notorious for the large amounts of mess they make, often leading to calls for mass extermination!

Feral Pigeons have managed to infiltrate the Rock Dove population of Britain, with a drastic effect on the wild, pure Rock Dove. Feral Pigeons breed readily throughout the year, and, being more aggressive than the placid Rock Dove, they are now moving steadily northwards.

Coming in an incredible variety of different plumages, colours varying from white to black to brown to grey, with a mix of different combinations and patterns, the Feral Pigeon is nevertheless an unmistakable bird.

▶ This pigeon is one of the many variants found within the Feral Pigeon group. It is dark grey on the head, underparts and most of the upperparts. The uppertail is greyish and the rump white. The wings are barred dark and light grey. The eye is brownish-orange, with a white orbital ring. The legs and feet are pink.

▲ This bird is fairly close to the true Rock Dove. The blue-grey head merges into a metallic green sheen on the neck and the upper breast shows a purple gloss, contrasting with the greens and greys.

One of the most familiar sights in urban areas is that of a flock of Feral Pigeons gliding effortlessly over rooftops. The Feral Pigeon has a very easy, sailing flight, which is interspersed with slow, deliberate flaps and glides.

The sheer range of variation amongst Feral Pigeons is well illustrated here. The bird on the right shows some of the familiar 'chequering' on the wing, common to many Feral Pigeons, while the two birds on the left show the typical pale end of the scale, one being washed pinky-buff, while the other is paler still, almost pure white.

TURTLE DOVE
(26-29CM, 10-11IN)

T HE TURTLE DOVE IS A SMALL, SLENDER-LOOKING BIRD which is common across much of Europe during the summer months. Usually shy, it is found in woodlands, plantations and bushy hedgerows, but can be a visitor to large urban or rural gardens.

Turtle Doves usually arrive in Britain in mid April to early May, and stay in this country until mid to late September. At coastal sites in spring, hundreds can be encountered in a day, pouring in off the sea after their long migration from Africa.

Small-headed, with a slim body, long graduated tail and variegated plumage, the Turtle Dove is a delightful and distinctive little pigeon. Its soft and deep purring 'rrooorrr rrooorrr' song, often repeated for long periods, can be heard throughout late spring and early summer.

The head shows a delicate grey cast to the forehead, crown and nape, with a buffy-pink face. The neck shows several black bars with distinctive white edges. The mantle is tawny-brown, fading into a slate-grey rump and dark closed tail. The wing is rich chestnut with bold black feather centres, with the exception of a grey forewing and dark flight feathers.

The breast is pinkish-grey, fading to off-white or pale grey on the flanks and belly. The bill is pale-based and black-tipped. The eye is orange with a rich red orbital ring. The legs and feet are reddish-brown.

▲ When seen from below, the appearance of the Turtle Dove is very striking. The pale grey head and pink breast contrast with the dark grey and black underwings and the black and white tail. Notice again the black central tail feathers.

▶ The upperwing view of the Turtle Dove is very different from the underwing view. Grey, chestnut, black and white all clash in a flurry of colour. Particularly striking is the tail pattern – grey central feathers, fading to brown and then black, with broad white tips except on the central tail feathers.

◀ Juvenile Turtle Doves show a buffy head, no neck bars and more subdued duller brown ring markings. The wings also show less grey, while the breast is far more buffy than the adult's.

A FAMILIAR SUMMER VISITOR, Cuckoos are found across Europe from early April to mid September. Cuckoos frequent a wide variety of habitats, from woods to reedbeds, coastal dunes to moorland.

The Cuckoo's long tail and pointed wings sometimes lead to confusion with certain birds of prey such as falcons or hawks. The fast flight is also rather raptor-like. However, once its famous call is heard, there is no identification problem.

The Cuckoo has developed a very sneaky way of raising its young. The female will lay her eggs in other birds' nests, garden species such as the Dunnock and Robin always being favourites, and 'allow' the new parent the pleasure of raising a huge youngster, which slowly but surely 'evicts' the other eggs and nestlings.

Upon arrival, the male's far-carrying familiar 'cuc-coo' call resounds around the country. Males will also utter a stuttering 'cuc-cuc-coo' and, when agitated, a gargling, laughing 'gug, gug, gug, gug'.

Cuckoos fly in a very direct manner, usually fairly low to the ground, but they are also very capable of gliding some distance. On landing, the wings droop and the tail is raised.

A male Cuckoo shows an entirely silvery-grey head and upper breast. The mantle and most of the wings are very slightly darker and the flight feathers are very dark grey. The rump is the same shade of grey as the head and breast, while the tail is again dark, but with white notches on each tail feather, complete with an obvious white tail tip. The underparts from lower breast to the undertail-coverts are barred black and white. The bill is orange-yellow at the base with a dark tip, and the eye is orange with a thin yellow orbital ring. The legs and feet are orange-yellow.

The distinctive rufous-brown phase of the female Cuckoo is scarcer than the more familiar grey birds, but not so rare as some books would have you believe. The entire upperparts of the bird, except for the primaries, show a rufous ground colour with fine black barring from head to tail. The underside is barred black and white from chin to tail, with a cinnamon wash on the cheeks, breast sides and the underside of the tail. The bare parts are as the male's. In flight, the colour of the upperparts, coupled with the shape of the bird, presents a very striking sight.

Grey-phase female Cuckoos are identical to the male except for a distinct brownish wash on the breast and browner-looking flight feathers.

Juvenile Cuckoos, once fledged, still take advantage of the surrogate parents' hospitality. Their plumage appears very scaly above – a dark grey ground colour on the head and back, with white edges to the feathers. The wings are similar, but show more brown flecks. The throat and upper breast are marked with dense black and white bars, which are more widely spaced from the lower breast to the tail.

BARN OWL
(33-36CM, 12½-14IN)

THE BARN OWL IS A MEDIUM-SIZED OWL with a heart-shaped face, yellow-brown and grey upperparts and white underparts (buff on Continental birds), with a wavering flight, piercing, shrieking call and overall 'ghostly' appearance. It is instantly recognisable to all and is a very beautiful bird indeed.

Barn Owls favour open countryside, hunting along field edges, scrubby areas, dykes, ditches and woodland edges. They nest in barns, special owl boxes and hollow trees.

They are found throughout western and northeastern Europe, and in Britain after a sharp decline due to all manner of reasons – poisoning, trapping, dubious building practices and an increase in traffic casualties – the Barn Owl is, slowly but surely, rising in numbers.

This is a typical view of a Barn Owl – a ghostly apparition drifting away over the fields showing broad, rounded wings, yellow-buff upperparts, barred tail and white underwings. During the winter months and breeding season, Barn Owls can often be seen in daylight hours.

The *alba* race of Barn Owl found in western Europe has a white heart-shaped face with an indistinct thin, grey border and striking black eyes. The rest of the head, the mantle and most of the wing are rich yellowy-orange and mixed with greys, blacks and browns. The short tail shows three grey bars. The underparts are white, flecked on the flanks and thighs. The longish legs are feathered white, while the feet are pinkish and have black claws.

This is a typical chick's eye view of a parent bird approaching, with a suitably juicy morsel. The adult birds will work extremely hard to keep their offspring nourished and during the breeding season no small mammal is safe!

An occasional visitor into Britain is the central and eastern and northeastern European dark-breasted race *guttata*. Structurally identical, these Barn Owls are considerably darker on the upperparts than the nominate race *alba*, appearing a lot greyer on the mantle and wings. The heart on the face is washed greyish-brown, while the underparts are a rich buffy-brown with neat dark spotting from breast to belly.

Waiting in a suitable roof space for the return of an adult laden with food, the young Barn Owls resemble a solemn line of choristers. The down of their younger days is soon lost – bar a few tufts on the head – as the familiar plumage begins to moult through. Adult Barn Owls will have done a good job if they manage to fledge three juveniles – they need good weather, sympathetic landowners and a large supply of mammals to succeed.

LITTLE OWL
(21-23CM, 8-9IN)

THE LITTLE OWL IS A DISTINCTIVE, BOLDLY SPOTTED SMALL OWL that is found widely throughout Europe, with the exception of northernmost countries. Little Owls were introduced into parts of Britain towards the end of the 19th century. They spread to colonize most of England and Wales, but they are rather sedentary and are still largely absent from Scotland. In Britain its favoured habitats include parkland, farmland and gardens in rural, suburban or urban areas.

Little Owls can be encountered at almost any time, being generally more active in daylight hours than other owls. The prime times are, however, dawn and dusk, as would be expected.

The general shape of a Little Owl is that of a small, squat, broad-headed bird with a short tail. The plumage is the same for both sexes.

When seen from the side, the Little Owl shows a very broad, dusky white supercilium which extends from the base of the bill to the back of the nape. Another white stripe extends from below the bill upwards over the rear cheek to join with the supercilium. The rest of the upperparts are chocolate-brown in colour, boldly spotted and streaked with white. The short square tail is barred brown and white. The underparts from throat to rear belly are brown with bold white marks. The undertail is white.

When seen head-on, the striking, fierce, bright yellow eyes of the Little Owl appear particularly obvious, while the white stripes around the head give the bird an almost 'grumpy' look. The small hooked bill is dull flesh-coloured and the legs are feathered white with dull grey 'toes' and black talons.

In flight, the rounded wings of the Little Owl are obvious. They have a fast undulating flight, a little reminiscent of Mistle Thrush or woodpeckers, and can often be encountered flying between telegraph poles. They are capable of easy movement on the ground when searching for food.

As the sun sets, Little Owls become very active. Listen out for the feline-like 'kee-uw' call or, during the breeding season, a variety of canine-like yelps and even barks. They will often call from old buildings, chimney pots or their beloved telegraph poles.

Juvenile Little Owls show vaguely similar plumage patterns to adults, although the face pattern in particular is a little more subdued, while the overall plumage appears less contrasty and has a 'downy look' to it. The white splodges are less well defined, particularly on the head and upperparts, while the underparts appear more streaked than spotted.

TAWNY OWL
(36-40CM, 14-16IN)

THE TAWNY OWL IS A MAINLY NOCTURNAL, MEDIUM-SIZED PLUMP OWL that is commonly found throughout most of Europe. It can be found, mainly by voice, in cities, towns and open countryside, in woodlands, parkland and gardens.

Tawny Owls are seldom encountered in daylight hours, being seen only by chance or at a known roost tree. If they are seen, it is much more likely to be at the very depths of dusk or perhaps caught in headlights perched on roadside signs or posts.

Large-headed birds, Tawny Owls are also decidedly stocky, despite the reasonably long body. Their highly complex, intricate plumage varies from reddish-brown to grey-brown, although in Britain the 'red' phase is by far the commonest form that will be seen.

You are far more likely to hear a Tawny Owl than you are to see one. The oh-so-familiar 'too-wit, too-woo' call is *the* Owl call. In actual fact this instantly familiar call is closer to a long drawn out 'oo-ooo-hooo'. Listen for the penetrating 'kee-wick' call as well.

The large head, which lacks any tufts, shows a warm reddish-brown area around the eye and cheeks, except for two white 'c' shapes (one 'reversed') near the bill. The crown is brown with darker vermiculations and two white lines which extend to the hindneck. The remainder of the head is a more tawny brown. The upperparts are entirely warm brown, with all manner of black vermiculations, bars, fringes and edges. The wings show two or three bold white areas of marks, on the scapulars and coverts. The outermost flight feathers are more yellowy-brown. The tail is finely barred. The underparts have a warm brown wash to the breast and flanks, contrasting with blackish notches from breast to undertail. The large eyes are black, with a tiny silvery bill and feathered feet.

In flight the Tawny Owl appears round-winged, square-tailed, dark above and pale below. The face looks strikingly pale, accentuated by the black eyes and dark crown. The rows of white spots on the wing and the barred tail are discernible, and check for the barred tail. The underwing appears barred blackish and white on the flight feathers with a whiter forewing, with dark smudges.

'Grey' Tawny Owls are rarely seen in Britain. The plumage patterns are identical to the more familiar 'red' birds, except that all the rich rufous-brown colouration is replaced by cold ashy-grey tones. The breast lacks any sort of wash, just black and greyish-white. The bare parts are the same as on 'red' birds.

Juvenile Tawny Owls have a very striking inquisitive look about them, as large black eyes peer out from a mass of down and feathers.

KINGFISHER
(15-16CM, 5½-6IN)

THE KINGFISHER IS AN INSTANTLY RECOGNISABLE BIRD, commonly found across most of Britain and Europe. Despite favouring more secluded waterways, Kingfishers can be seen on almost any river, stream, fishing-lake, gravel pit or garden pond. However, despite the spectacular colour scheme, they can be notoriously difficult to see.

Kingfishers can spend a great deal of time sitting on a half-hidden branch over a quiet river, waiting for a chance to prove their prowess. At other times, however, Kingfishers can be noisy beasts, whether squabbling over territory or planning intimate liaisons, and their piercing, piping call can be heard several hundred metres away.

Kingfishers have a large head, a thick-based dagger-like bill, short wings and a short tail. The bill is black on the male (sometimes with deep red at the base), while the female has a blackish bill with an extensive orangey-red tone to the lower mandible. Both sexes have black eyes and red feet.

Kingfishers really speed past you as they fly by, an amazing sight when first encountered, a whoosh of colour. The blue of the upperparts contrasts with the darker, greener-looking flight feathers, while the white neck patch and orange cheeks and underparts can all be seen if you are lucky enough.

This adult male's head pattern is a mix of brilliant blue, rich orange and white. The forehead to the rear nape is a deep turquoise-blue, while the lores and cheeks are orange, and the throat and neck patch are white. The upperparts are bright turquoise, with a slightly darker tail. The wings are a slightly darker blue, with a small amount of white spotting. The underparts are wholly orangey-red.

Kingfishers nest along 'softish' river banks, choosing somewhere that will give them every opportunity to have an easy dig as they excavate the nest tunnel. They will lay between five and seven eggs, and usually have two broods in a season.

The classic view of a Kingfisher, plunging into the water in search of a suitable fish dish. Wings held back, neck extended, bill thrusting forward and then SPLASH!!

Juvenile Kingfishers do not have quite the same technicolor plumage as the adults, but they are still pretty snazzy. The head and wings are more green than blue, and the orange of the face and underparts is a little more subdued. The mantle and tail are quite bright blue, though. The bill and eye are black and the feet orangey. When perched, Kingfishers, if nervous, will bob their heads and flick their tails before whirring off into the distance at great speed.

GREEN WOODPECKER

(30-33CM, 11½-13IN)

THE GREEN WOODPECKER IS A LARGE, STURDY WOODPECKER found throughout most of Europe, with the exception of Ireland and northern Scandinavia. Green Woodpeckers can be seen in almost any wooded area, from cities to the open countryside.

Although primarily a tree-loving species, Green Woodpeckers are very much at home on the ground, hopping about on short-turf areas, such as your lawn, in search of insects.

A long, robust-looking bird, the Green Woodpecker has a powerful, sharp silvery bill with a black tip, large grey feet (with two toes forward, two back) and a short, spiky tail which the bird uses as a pseudo-crampon for supporting itself on the tree trunk.

◀ The Green Woodpecker has a distinctive head pattern: red on the forehead extending to the nape, black around the eye, a black line bordering the forehead and a black moustache. On the female the moustache is wholly black. The cheeks and throat are pale greyish-white to pale yellow. The mantle and wings are olivey-green, although the outer wing feathers are notched dark grey and white.

▶ On a male Green Woodpecker the centre of the moustache is red. Both sexes have a dull yellow rump and uppertail-coverts, while the graduated tail is brown-grey with darker bars. The underparts are greyish-white except for subtle black notches on the undertail-coverts. The eye is white and the undertail itself is blackish.

▼ Like all woodpeckers, Green Woodpeckers fly in a characteristic undulating manner. Deep beats are followed by clo   sed wings as the bird progresses, rather rapidly, from site to site. As it approaches a tree, the bird will come in with a huge upward sweep before clamping itself to the nearest trunk.

▲ Often when treeward-bound, but also when perched, the distinctive, far-carrying, ringing 'laugh' or 'yaffle' of the Green Woodpecker will be heard. The yellow rump will be particularly obvious as the bird flies, contrasting with the comparatively dark wings and tail.

▼ When feeding on the ground, Green Woodpeckers will alternate between probing for food (usually ants) and looking up, cautiously, to survey the situation. The long, sticky tongue may be more easily seen as the woodpecker feeds on the ground, before bounding off clumsily to the next ant-hill.

▼ Juveniles show certain significant differences in their plumage. The white face has fine black streaking and lacks the moustache, which becomes more visible as the bird gets older. The red crown is also streaked black. The upperparts are paler green, while the underparts are pale with marked streaking and barring, particularly on the breast.

GREAT SPOTTED WOODPECKERS ARE EASILY IDENTIFIED, thanks mainly to their pied plumage, but also size, plumage patterns and call.

This is a medium-sized 'pied' woodpecker which is widespread across Europe, except for Ireland and the very northern edge of Scandinavia. It is becoming an ever more common bird in gardens, particularly in the winter, when it will take full advantage of rind, suet and nuts left out for it.

Away from gardens, Great Spotted Woodpeckers are easily found in most woods and copses, in any suitable area, in villages, the great outdoors and cities.

Great Spotted Woodpeckers, although smaller than Green Woodpeckers, share the same woodpecker characteristics – thick pointed bill, graduated tail, 'reverse' toes and undulating flight.

Great Spotted Woodpeckers have an explosive, excited 'chick' call, mainly given in flight, and in spring have a very loud 'drum'-tapping on a tree at a rapid rate, faster than any other woodpecker.

The male (right) has a pale greyish area on the lores, and a jet-black crown, moustache and near vertical line behind the ear-coverts. This contrasts markedly with the white cheeks and neck patch. On the rear crown is a striking red patch, which is absent on the female (above). The mantle, rump, uppertail-coverts and tail are black, except for the outer tail feathers which are white, with fine black notches. Notice the longer central tail feathers. The wings show bold white shoulder patches and white spotting across the otherwise black remainder. The black flight feathers are tipped white. The underparts are white, with a black shoulder patch and red vent area. There is often a buffy wash across the breast.The short, dagger bill is deep silvery-grey in colour, the eye is black and the feet are dark grey.

In flight, the Great Spotted Woodpecker is particularly striking. The bold white shoulder patches and white spotting on the wings really stand out. The red vent is visible and the black and white face is obvious. From below, the forewing is startlingly white, compared with the spotted hindwing and primaries. The black and white graduated tail contrasts clearly with the red vent. Just like the Green, the Great Spotted Woodpecker has a particularly undulating flight.

The young Great Spotted Woodpecker is basically like the adult, but shows subtle differences. The crown is entirely red, unlike that of the parents. The black moustache is thinner and often streaked white. The white of the shoulders and flanks often shows some fine black streaking, while the underparts never appear quite so snowy white as on the adults, and the red on the vent is decidedly duller.

LESSER SPOTTED WOODPECKER
(14-15CM, 5½-6IN)

THE LESSER SPOTTED WOODPECKER IS A CHARACTERISTIC, SMALL, ROUND WOODPECKER which, although shy and hard to see, is common across most of Europe except for northern Britain and Ireland.

Lesser Spotted Woodpeckers can be seen in parks and woods, even in cities, and are easiest to see in late winter and very early spring when leaves have yet to emerge and the woodpeckers become quite vocal.

Europe's smallest woodpecker, only the size of a House Sparrow, the Lesser Spotted Woodpecker has a neat pied appearance with a small sharp bill, and the familiar graduated tail and reverse-toe woodpecker combination. When seen from behind as it clings to a trunk or small branch, it is best identified by its very small size and neatly barred back.

A male Lesser Spotted Woodpecker (left) shows a neat red crown, compared with the whitish crown on the female (above). This is the only major plumage difference between the two sexes. The face of the adult is largely white except for a thin black borderline to the crown, and a broad black moustache which extends slightly onto the cheek before curving up onto the side of the throat. The nape shows a narrow black line, joining the rear crown to the mantle. The back is barred black and white, fading to a black rump and central tail. The outer tail feathers and wings are barred black and white. The underparts are off-white, with some light flecking on the flanks. There is no red on the underparts. The short sharp bill is silvery-grey, the eye black and the feet and legs blackish.

Lesser Spotted Woodpeckers have a typically undulating flight, and, when seen from above, the barring on the back and wings is very striking. From below, the white on the forewings is obvious, as is the barred underwing. The underparts may appear buffy.

Lesser Spotted Woodpeckers have a piercing, almost Kestrel-like 'kee kee kee' call, often heard in early spring. They also have a quiet 'kick' call note, but this is rarely heard. The 'drum' is longer and more constant than that of Great Spotted Woodpeckers. They are also able to feed on smaller branches and twigs than the bulkier woodpeckers.

A juvenile 'Lesser Spot' superficially resembles the adult, except for a whitish forehead, sometimes flecked black, spottier flanks and a buffy wash to the face. This juvenile male shows some red on the rear crown.

SWIFT
(16-17CM, 6½IN)

THE SWIFT IS A FAMILIAR, DARK TORPEDO OF A BIRD, common in the summer months across the whole of Europe. Arriving in late April from African wintering quarters, the Swift will stay in Europe until late August or early September, before undertaking its massive journey back southwards.

Swifts can literally be seen anywhere, whatever the area – city, town, or hamlet. So long as it has eaves to nest under, the Swift is happy. If the weather suddenly becomes very rainy and windy, however, all the Swifts will immediately move away to more hospitable areas, sometimes covering very long distances before returning after the storms have subsided.

With its scythe-shaped wings, short forked tail and generally dark brown plumage, it is a fairly unmistakable aerial feeder, seldom seen clinging to walls. Being such adept and competent fliers, Swifts can display a wide variety of shapes when airborne.

The plumage of the Swift is almost wholly dark brown, blackish, when seen from a distance, except for a small whitish throat patch. With close views, a subtle contrast will become apparent between the blackish forewings and the dark grey remainder of the wing. With exceptionally close views, the tiny black bill can be seen, as can the black eye.

The sooty-brown tones of the upperparts are easily seen when Swifts fly close by. The trailing edge of the wing may appear slightly paler in some lights. Swifts are very versatile when airborne and, as already mentioned, they can 'change shape' with great ease. In fast flight, the tail is closed, looking very pointed, and the wings are held well swept back, with very quick beats. When soaring, the wings are held further forward and the tail is open, giving an altogether different, stubby silhouette.

One of the most familiar summertime sights, particularly in the mid-evening, is of parties of screaming Swifts careering over rooftops, sometimes below head height but always in control. They seem to delight in their unrivalled skill in the air and almost shriek their pleasure to let everyone know just how good they are.

When Swifts make 'landfall' they are immediately transformed from majestic flying machines to ungainly, sad-looking birds, grimly holding onto a wall as if their life depended on it. Doubtless, this is *the* reason why they spend so much of their lives in the air.

Juvenile Swifts are hard to separate from adults when seen on the wing. At closer ranges, however, they will appear browner and more scaly, with white flecks on the forehead and a more extensive white bib on the throat.

SWALLOW
(16-22CM, 6½-8½IN)

T HE SWALLOW IS UNDOUBTEDLY ONE OF THE HERALDS OF SPRING, as it arrives back into the whole of the European continent in early April. As with other mainly aerial feeders, Swallows can be encountered in any situation, from coastal marshes to urban streets.

Always active, the Swallow benefits from its association with man, frequently nesting on the side of houses, outbuildings or barns. When not flying to and from their mud nests, Swallows will sit on telegraph wires, chattering cheerfully, while rearranging their feathers, before floating off on another feeding sortie, either singly or in small groups.

Swallows can be separated from Swifts and House Martins by their size, more pointed wings, the manner of flight and obvious tail length.

The Swallow's call is a tinkling and merry 'vit vit' or an occasional 'splee-plink' Their song is a strong, clear, rather fast and prolonged twittering warble.

▶ Swallows have a small, rounded, deep glossy blue head, and a chestnut face. The blue extends to the rump, shoulders and upper breast, forming a neat band. The wings are blackish in appearance. The underparts from the lower breast to undertail-coverts are cream – and are a variable feature. The bill, eye and legs are all black.

▼ The male Swallow is identical to the female in every respect, except for longer tail streamers. The bird depicted here has a particularly long set of tail streamers, and so is sexed as a male. This feature becomes obsolete as the season progresses and the feathers wear.

In the autumn, Swallows start to congregate on telegraph wires, TV aerials and roadside wires in preparation for their journey south of the Sahara. The noise can be overwhelming on occasion as they excitedly chatter 'vit, vit, vit, vit' to each other. Many hundreds, even thousands, will stop off at coastal reedbeds in the autumn, to feed on the profusion of airborne insects and to roost, in safety, amongst the reeds.

In flight, the most obvious feature will be the deeply forked tail which gives the Swallow such a graceful air. Notice also the white of the forewing, the pale belly and the white spots on the underside of the tail, which are particularly obvious as the Swallow fans its tail when banking. Otherwise the upperparts look wholly dark.

Juvenile Swallows are a duller blue on the upperparts, and the face is more dirty orange than chestnut. The breast band appears quite smudged, and the underparts are buffier than the adult's. The tail streamers are very short. Very young fledglings may show one or two flecks of down on the head and a strong yellow gape.

HOUSE MARTIN
(12-13CM, 4½-5IN)

THE HOUSE MARTIN IS A FAMILIAR SUMMER VISITOR that is seen across the whole of Europe. As with their close cousin the Swallow, House Martins can be found anywhere, making ample use of human habitation for nesting sites.

House Martins spend a good deal of time in the air, but, unlike Swifts and Swallows, they can be seen on the ground, collecting wet mud for use as nesting material.

Clearly smaller and more compact than the Swallow, House Martins, when seen well, should be unmistakable. When flying at lower levels, their short, hard calls can easily be heard. In fine weather, however, they frequently hunt for insects very high in the air, often mixing with Swifts, and at such times their smaller size is generally obvious.

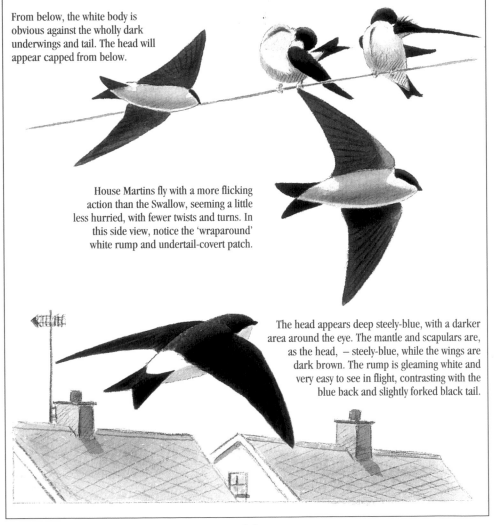

From below, the white body is obvious against the wholly dark underwings and tail. The head will appear capped from below.

House Martins fly with a more flicking action than the Swallow, seeming a little less hurried, with fewer twists and turns. In this side view, notice the 'wraparound' white rump and undertail-covert patch.

The head appears deep steely-blue, with a darker area around the eye. The mantle and scapulars are, as the head, – steely-blue, while the wings are dark brown. The rump is gleaming white and very easy to see in flight, contrasting with the blue back and slightly forked black tail.

Favourite of all nest sites for House Martins is the 'under the eaves' position. They will return year after year to the same building, and now the familiar mud nests have been 'copied' by nestbox manufacturers, who have produced a synthetic equivalent.

The juvenile House Martin (left) is a dull version of the adult (right), appearing browner on the crown and cheeks, duller blue on the mantle and scapulars, and browner on the wings, with white tips to the tertials and primaries. The underparts show a dingy wash to the throat, upper breast and flanks. Very young fledglings show a yellowy gape at the base of the bill.

When seen collecting mud, House Martins appear quite content, so long as they do not have to walk too much. They actually spend very little time on the ground, being far happier in the air. The bare parts can be seen well when a House Martin makes touchdown. The small bill is black, as is the eye, while the legs are feathered white, and the feet are pale.

GREY WAGTAIL
(18-20CM, 7-8IN)

THE GREY WAGTAIL IS A CHARACTERISTIC, SLIM, COLOURFUL BIRD that is widespread across much of Britain and southern Europe, but absent from most of northern Europe. During the breeding season it is found only alongside running water, but in the period outside the nesting season, it can be seen around lakes, ponds and even city centre rooftops.

Grey Wagtails are very active birds, constantly on the move, as they dip from rock to rock in search of food, with their long tail always wagging.

The general slim lines, the very lengthy tail and the overall plumage make the Grey Wagtail an easy bird to identify. You should have little problem eliminating the Pied or Yellow Wagtail.

The female (right) differs from the male (below) only in the absence of the black throat patch, and has less yellow on the breast. Like the male, the female has a closed wing which is black with a white edging, a yellow rump and a black tail with white outer tail feathers. On both, the thin bill is black, the eye is also black and the legs and feet are fleshy-pink.

In his summer outfit, the male is a handsome bird with a slate-grey head and back, a white supercilium and a white moustache, bordering an obvious black throat patch. The breast and vent are flushed strongly with yellow. The remaining underparts are off-white.

In flight, the long black and white tail will be immediately apparent. Look on the upperwing for an obvious white bar, dividing the grey of the forewing and the black of the hindwing. The yellow of the rump will be easily seen. Listen out too for the distinctive call note, a ringing metallic 'st-it' or 'tzit'.

In the winter months, the male Grey Wagtail (left) loses his black throat and his yellow breast becomes very pale, while the yellow on the female becomes more buffy-white.

Juvenile Grey Wagtails look like females, although they are buffier on the upperparts, and faint white wingbars may be apparent. The rump and undertail will be greener than on the adults, while the buffy underparts show some faint streaks on the breast.

PIED WAGTAIL
(17-18CM, 6½-7IN)

T HE PIED WAGTAIL IS AN INSTANTLY FAMILIAR BIRD which is common throughout Europe. It can be seen in any location, from coastal marshes to city gardens. Pied Wagtails are very approachable, having little fear of man, and feed and breed quite happily in close proximity to human activity.

A busy bird, the Pied Wagtail provides hours of pleasure, darting, flicking, running and chasing off in search of insects. During the winter, large flocks of Pied Wagtails have become an increasingly familiar sight around city streets as they fly to roost amongst trees at dusk.

Their plumage is ultra distinctive and no other bird shows these plumage characters coupled with a thin bill, rounded body and long wagging tail.

On the Continent the 'Pied Wagtail' is replaced by the 'White Wagtail'. Both are part of the same species, but Pieds are found in Britain, Ireland and nearby Continental coasts, while the White Wagtail is common across the rest of Europe.

In the summer there is little difference between the sexes. The male (left) has entirely black upperparts, except for a white forehead and white cheeks. The black wings have prominent white edgings to all the main feather groups – particularly obvious are the two 'covert' wing bars. The underparts show a large black bib, which extends from the base of the bill to the upper breast. The rest of the underparts are white. The female (below) differs only in having a greyer back and less black on the head and breast. The bill, eyes and legs are all black.

In flight, the black and white of the Pied Wagtail is very striking. The two white wing bars contrast markedly with the dark wings and wholly black upperparts. The white outer tail feathers and the gleaming white underparts are obvious. It has an awkward undulating flight, often calling with a high-pitched 'chiz-zik' or 'seel-vit'.

The male White Wagtail (left) differs from the Pied Wagtail in showing an entirely pale silvery-grey back and rump, which contrast strongly against the black of the nape. Female Whites resemble the male, except for a duller back and less black on the head.

A first-winter Pied Wagtail basically looks identical to a female, and in some cases they are indistinguishable. Usually it can be told by a browner tint to the back and a dark grey (not black) crown.

The juvenile is a grey and brown version of the adult. The head is greyish-brown, with a white 'ear-covert frame' and buff-white throat. The upperparts are browny-grey, with a dark greyish rump and black tail. The underparts are buff with an indistinct black bib and grey shoulders.

WAXWING
(17-18CM, 6¹/₂-7IN)

THE WAXWING IS ONE OF THE MOST ATTRACTIVE GARDEN BIRDS. There is no other bird which even vaguely resembles it and, although it breeds only in north and northeastern Europe, it appears in western Europe every autumn and winter, sometimes in large numbers.

Waxwings are portly birds with longish wings and tail. They can be encountered in gardens singly or very occasionally in a large flock. Both sexes show a distinctive, bold pinky-brown crest on the head. The forehead and bill base are chestnut. The remainder of the head, mantle and scapulars are pinkish-brown. The wings are a complex pattern of grey tertials, black and white secondaries, with white and yellow 'v' tips to the primaries. The rump and uppertail-coverts are pearly grey, extending to the tail, which has a broad yellow tip. The breast is pinkish, fading on the belly and flanks. The undertail-coverts are chestnut. The hefty-looking bill is silvery-based with a blackish tip. The eye is black and the legs and feet are silvery.

▼ The male Waxwing shows a distinctive head crest and a jet-black eye mask and bib. The waxy red tips on the end of the secondaries are quite marked. These give the bird their name.

▲ Females are similar to the male, but do show some differences. Most obvious is the slightly smaller, shorter crest and the less well-defined sootier bib. The waxy red tips on the secondaries are less obvious, as is the yellow tail band and primary tips.

90

Waxwing flocks move very quickly, and despite their dumpy appearance they are very adept fliers. The shape in flight is reminiscent of Starlings, but Waxwings actually look thinner. If you do encounter a group of Waxwings, listen out for their delightfully gentle, but ringing, easy on the ear call – a whistling 'sirrr'.

When seen from underneath, the black bib and cinnamon breast are noticeable, as are the gleaming white underwings, rufous vent and grey-yellow tail. The upperwing shows dark flight feathers, and the grey on the rump is obvious. The dark russet crown can be seen, as well as the black and yellow tip to the rather short tail.

Young Waxwings, while still utterly attractive, do lack some of the panache of the adults. The crest is shorter, the plumage is browner, the black is restricted to the chin, the waxy red tips are almost always missing and the yellow on the tail is not very bright. This group is seen in a typical scene, feeding avidly on any number of red cotoneaster berries!

WREN
(9-10CM, 3½-4IN)

THE WREN, DESPITE ITS TINY SIZE, IS JUST FULL OF ENERGY! Always busy, always actively seeking food or shelter, this species is one of the commonest garden birds in the region.

Wrens can be found in all manner of different habitats away from your garden. They are particularly abundant in woodland, and are also commonly found around scrubby areas, farmland, reedbeds and even (in the Northern and Western islands of Scotland) on cliffs.

Wrens have a very distinctive appearance – tiny with a short cocked tail, and a plumage that is a rich mix of brown shades. The decurved bill is silvery-black, with a pale base, and the legs and feet are fleshy-pink. They are frequently secretive in nature, furtively seeking food in the cover of tangled undergrowth, and can be hard to see. Only a little patience is required, however, before they reveal themselves on the topmost twig of a bush or on top of a garden fence.

One of the most familiar views of a Wren is this – perched on a bramble delivering its rapid-fire, scatter-gun scold 'cherr, cherr, cherr' or its explosive loud song, high-pitched and trilling. The stumpy tail is held cocked to the sky, revealing buffy grey and black markings on the undertail.

The slightly decurved bill is opened to reveal a glowing yellowy gape.

The Wren, despite its initial look of being 'just brown', is, in fact, a delicately marked bird full of tone. The head is a rich dark brown, with a broad creamy-white supercilium, black eyestripe and streaked grey-black ear-coverts. The mantle and forewings are also rich dark brown, fading to a rich rufous on the rump and tail, which is thinly barred with black. The tertials are rufous, barred black, and the secondary feathers are barred greyish-white and black. The wing tips are black. The underparts are greyish merging into brownish-yellow, and are prominently barred black on the flanks.

In flight, Wrens are a flurry of activity. Wings whir furiously, too fast to see the flaps, as they hurry from cover to cover. Often all you will notice is a tiny brown bird whizz past and then you will hear the scolding calls again, or a snatch of the surprisingly loud song, but the bird itself will often be invisible inside the cover into which it has retreated.

Wrens build a perfect round nest, usually of dead grass and vegetation, lined with moss or discarded feathers, often located amongst brambles to lessen the chance of predators. Parents make hundreds of journeys to satisfy the hunger of their clamouring brood, which after a week or two venture out of the nest to beg for food.

During the depths of winter, Wrens will utilise nest boxes as a communal roost site. As dusk approaches, little brown shapes can be seen rushing towards such a box, and a peek inside could reveal up to thirty or more birds tightly packed together for warmth. This helps them to survive the often cold nights during which many might otherwise perish.

DUNNOCK
(14-15CM, 5½-6IN)

THE DUNNOCK IS A RATHER DARK, FAIRLY SHY AND RETIRING GARDEN BIRD which is very common throughout Europe. Its favoured habitat is anything which is vaguely dense and scrubby, from gorse bushes on coasts to bramble clumps in woods or a simple garden hedge.

Despite their somewhat drab plumage, Dunnocks are rather an endearing species. They have a lovely intimate courtship display and a sweet warbling song. The Dunnock will often be seen darting out of cover, creeping furtively over a lawn, before darting back into the hedge. Aside from the high-pitched, clear and very resonant "sissisisis" song, the slightly less impressive, but still strong, call note 'seeh' is usually heard well before a Dunnock appears. The Dunnock's older name of 'Hedge Sparrow' reveals its liking for such habitats. This name is still used by many country folk today, although the bird is in no way related to the sparrows.

The adult shows a dull grey head and breast, with dark brown on the crown and on the ear-coverts, which can also show very faint white flecking. The mantle and wings are very warm brown, almost rufous, with quite bold blackish streaks. The wings may show a faint white wing bar. The rump is greyer and unstreaked, while the tail is dark brown. The flanks often show warmer russet tones, but also show strong dark streaks. The vent and undertail are off-white to buff. The thick bill is black, the eye dark brown and the legs and feet are dull pink.

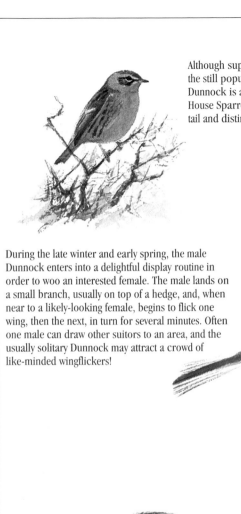

Although superficially resembling a sparrow (hence the still popular name, Hedge Sparrow), the Dunnock is a more slender, less dumpy bird than the House Sparrow and has a thin bill, longish-looking tail and distinctive plumage.

During the late winter and early spring, the male Dunnock enters into a delightful display routine in order to woo an interested female. The male lands on a small branch, usually on top of a hedge, and, when near to a likely-looking female, begins to flick one wing, then the next, in turn for several minutes. Often one male can draw other suitors to an area, and the usually solitary Dunnock may attract a crowd of like-minded wingflickers!

Juveniles are duller than adults. The head is browner, and the upperparts lack the strong rufous tones, appearing plain brown with black streaks. The underparts show heavy black blotches on the breast extending to the flanks. The throat and undertail are whiter than the adults. The bill is silvery-grey.

ROBIN
(13-15CM, 5-6IN)

THE ROBIN IS WITHOUT DOUBT one of the most popular of all garden birds, and with its plump, rotund shape, red breast and the male's delightful song the Robin is also one of the garden's most easily recognisable species.

Robins breed throughout northern Europe, in a variety of woodlands, parks and gardens. In Britain, the autumn and winter sees the population of Robins swell as Continental birds move across the North Sea, sometimes in very large numbers.

Adult Robins are very easy to identify, their shape and plumage rendering them unmistakable. Although the juvenile does not have a red breast and is very spotty-looking, it does have the same distinctive shape and behaviour as adults and cannot really be mistaken for any other bird.

Surely a highlight for any garden is to be able to enjoy the beautiful warbling song of the Robin. Males seem to require little prompting to burst into song, and cold winter days are often enlivened by this red-breasted serenader. Robins love to pick an obvious perch from which to sing and they will dominate the dawn chorus!

A small, fat bird with, seemingly, no neck, the most obvious plumage feature is the bold orangey-red face and breast, bordered by grey. The rounded head is olive-brown on the crown and nape, as are the remainder of the upperparts. The wings are slightly darker brown, particularly on the primaries, and the tail often appears darker towards the tip. The underparts (from below the breast) are off-white, with a buff wash on the flanks. The bill is short, relatively fine and black. The eye is beady and black, while the legs and feet are light brown.

The Robin has a variety of familiar postures. Sometimes strutting, sometimes inquisitive, it is always on the move. Quick hops are punctuated by short stops, when the short wings droop, the tail cocks and the head is held slightly to one side. And then it is off again!

Robins are prone to short flights, often close to the ground. They move from perch to perch, darting onto the ground as they go, with little flicky flights.

Juvenile Robins are, in their own way, just as distinctive as the adult bird. The head, upperparts and breast are liberally speckled with dark-brown spotting and scalloping, which sometimes extends onto the flanks. A bold white eyering is easily seen. The remainder of the underparts are off-white. The bill is almost entirely black, except for a yellow gape. Eye and leg colour is the same as for the adult. The young Robin will moult into its red-breasted garb between June and September, and, just as the adult, will become progressively more tame as the breeding season fades to autumn.

THE FIELDFARE IS A LARGE, ROUND-LOOKING THRUSH which breeds in northern Europe and winters in large numbers across western Europe into Britain and Ireland. Small numbers do, however, breed in northern England and Scotland, and the Fieldfare is officially recognised in Britain as a rare breeding bird. Fieldfares like to breed along woodland edges, but on the wintering grounds they can be seen in almost any habitat, especially open arable fields, hedgerows and gardens.

In the winter, Fieldfares commonly associate with other thrushes, particularly Redwings, forming large roving flocks which can quickly devour a berry-filled hedge. If the winter weather is harsh, the Fieldfare seeks out food in sympathetic gardens, and is particularly fond of any windfall apples.

Reminiscent of the Mistle Thrush in size and shape, the Fieldfare is easily identified by its very distinctive maroon, grey, yellow, white and black plumage, and by its very harsh loud chattering 'chak, chak, chak' call. The sexes are similar, although with care subtle differences can be seen.

The male Fieldfare (below right) has a dove-grey head, with an indistinct white stripe above the eye, bold streaks on the crown and a black line around the cheeks. The mantle is a dull reddish-brown and the wings are similar, except for the dark-centred tertials and blackish primaries. The rump is grey, the tail black. The underparts show black arrowheads extending from the chin to the rear flanks, while the breast is washed yellowy-buff and the belly is white. The female Fieldfare (left) has finer black streaks on the crown, and is generally duller. The mantle and wings are browner, the head and rump and the breast are paler, showing fewer black arrowheads.

Often seen in large loose flocks, the wayward and erratic flight pattern of the Fieldfare is in itself a useful field point. On a close view the white on the underwing will be clearly seen. On the upperparts, the grey rump, black tail and black-speckled reddish hue are distinctive field features.

A characteristic view of a Fieldfare is to see them perched, back-on, with wings held drooped and tail slightly cocked. The contrast between the russet-coloured mantle, the grey rump and head and the blackish tail is immediately apparent.

A juvenile or first-winter Fieldfare is much duller than the adult. The back is brown, with dark feather centres and indistinct white edges. The underparts show a dingy buff wash, and the black markings on the underparts are less clearly defined.

BLACKBIRD
(24-27CM, 9½-11IN)

The BLACKBIRD IS ONE OF THE COMMONEST GARDEN BIRDS in northern Europe, frequenting a broad range of different habitats in every possible garden context.

Blackbirds show all the classic thrush characters and traits; they are quite stocky, with plump bodies, rounded heads, longish wings and longer tail. They hop, run and flick the wings and tail in various degrees of agitation. They also scurry under cover at the merest sign of danger.

In Britain the resident population is augmented in the autumn and winter, with the arrival of birds from the Continent.

The male Blackbird (left) is the only jet-black garden bird. The plumage, in spring, can take on a very glossy look but into the summer months the wings, in particular, can look brownish through wear. The bill and eyerings are bright orange-yellow and the legs and feet are dark grey. The eye itself is jet-black. An adult female Blackbird (right) comes in a couple of subtly different guises – a rufous and a grey-brown version, with much variability in between. The head and upperparts can vary from dark chocolate-brown to dark greyish-brown with an olive wash. The tail is black. The throat is pale with fine dark streaks. The heavily mottled breast colour varies from dark tawny-brown to greyish-brown. The mottled belly tends to be paler than the breast. The bill is blackish-brown, with a yellowy base, and they lack the bright yellow eyering of the male. The eye and leg colour are as in the male.

▶ A resident male Blackbird will make his intentions very clear from early January onwards, with any perch being suitable, be it a rooftop, television aerial, wall or treetop. The song is a delight, full of flutey notes delivered with breath-taking clarity. One of the garden's champion songsters.

◀ Once moulted from its juvenile plumage this first-winter male shows the signs of adulthood. The upperparts are dull black, contrasting with dark brown wings – but lacking the olive tones of the female. The underparts, except the throat, are blackish-brown. The face and throat are greyish-black, with fine white streaks. The eyering is yellow and the bill is black.

▶ Pictured in a characteristic tail-cocked pose, juvenile Blackbirds appear more rufous than the adult female, and they are certainly more heavily mottled, on both the upper- and the underparts. The throat is whitish, with very thin streaks. The wings show broad fringes and tips, all indicating the very fresh plumage.

◀ Blackbirds are one species in which albinism (partial or total) is relatively commonplace. Many people will have seen a Blackbird showing liberally scattered random white patches. Sometimes a male Blackbird exhibits a white breast gorget, like that of its hill-loving cousin, the Ring Ouzel (top left). An albino with this feature often shows other white markings, and wholly black wings – Ring Ouzels have grey edges to their wings, the gorget is very broad and they are fractionally smaller.

SONG THRUSH
(22-24CM, 8½-9½IN)

T HE SONG THRUSH IS A FAMILIAR GARDEN BIRD which is widespread and common throughout the whole of Europe. In Britain and Ireland, these are regular visitors to gardens of almost any size, parks, hedges and woodland. Further north, they are commonly found in most woodlands, particularly favouring damper plantations and forests.

Having declined somewhat dramatically in the mid 1980s in Britain, the number of Song Thrushes seen in the UK is now on the up once again. In the autumn, the British population is swollen by Continental migrants which stream into the country during September to November.

The Song Thrush is without doubt *the* garden thrush and should not really be confused with its spotty cousin, the Mistle Thrush. The Song Thrush is clearly a smaller, more compact bird compared with the big and bulky Mistle Thrush. The shorter tail of the Song Thrush is also easily noted, as is the lack of obvious pale wing fringes and the less boldly marked breast.

▶ The bill, quite short but chunky, is dark-tipped with a warm brown base. To break open the protective shell of a snail, Song Thrushes dash them against a hard anvil, such as a stone or tree root. This is a common sight in winter when hard ground prevents a ready supply of worms.

◀ The adult's head is warm brown, with a faint buffy supercilium and very finely spotted patch on the cheeks. The mantle, rump and tail are all warm brown. The wings are brown and usually show two faint, thin off-white wing bars. The throat and belly are white, the breast and flanks are washed yellow, but most obvious is the bold black spotting which extends from throat to rear flanks.

▶ The Song Thrush's song is a clear, languid, flutey affair consisting of much mimicry and repeated phrases. Once one phrase has been sung three or four times, a new phrase will take over. In flight, the Song Thrush has a thin, but somehow far-carrying 'tsip' call. The alarm call is much like that of Blackbird and other thrushes, a loud 'chuck-chuck'.

▼ In flight, the spotting on the underparts will be seen immediately. Also, look for the orangey colour on the underwing, the only common thrush to show this pattern. The flight itself is quick and very direct.

▶ The juvenile Song Thrush is similar in plumage to the adult bird, but shows paler marks on the mantle, a more buffy wash to the underparts and smaller spots on the underside. Like the adult it has a black eye, complete with obvious orbital ring, and longish, pink legs and feet.

REDWING
(20-22CM, 8-9IN)

THE REDWING IS A SMALLISH, ATTRACTIVE THRUSH which can be found, in the breeding season, in northern Europe (Scandinavia and Iceland) and in the autumn and winter months throughout the rest of western Europe. Redwings breed in birch forests or bushes, old tree stumps and even hedgerow banks. On the wintering grounds, they behave in much the same way as the Fieldfare (in whose company they are often seen), feeding in berried trees, open fields and woodlands.

Just like the Fieldfare, Redwings will take advantage of a garden if the winter weather takes a turn for the worse. The need for keeping water unfrozen is paramount, and ensuring a good supply of food (windfall apples are a particular favourite) will doubtless encourage groups of Redwings into your garden.

The adult's head shows a dark brow, with a whitish supercilium and moustache, extending towards the cheeks. The ear-coverts are brownish, with very fine white streaks. The mantle, rump and tail are grey-brown, while the wing shows darker feather centres on marginally greyer overall colour tones. The flight feathers are darker still. The underparts are white, with slender streaks extending from bill base to flanks. The flanks usually show a vibrant red patch extending to the underwing.

Redwings superficially resemble the Song Thrush, but are smaller and darker, with a distinctive face and underpart pattern. The bill is dark and stubby with a yellow base. The eye is black and the legs and feet are fleshy-yellow.

In flight, the most obvious feature to be seen is the red on the flank and the underwing. With a close view, the head pattern and spots below should be seen. They often fly in disorderly flocks.

A very gregarious bird, the Redwing feeds in flocks and the birds often bound and hop along the ground in synchrony. If disturbed, they will fly off, with a thin, high-pitched 'tzipp' call.

A first-winter Redwing is basically identical to an adult bird, except for white or buffy tips to the greater coverts on the wings. The bare parts are the same as on the adults. When seen from behind, the Redwing will appear cold ashy-brown with the exception of the stripey head pattern and clearly fringed wing markings.

THE MISTLE THRUSH IS A LARGE, ROUND THRUSH, relatively common across northern Europe. A little more secretive than other thrushes, Mistle Thrushes can be found in gardens, coniferous woods and parkland, as well as farmland.

As with other thrush species that can be seen throughout Britain, their numbers can rise with the arrival of Continental Mistle Thrushes in the autumn and winter months.

Mistle Thrushes are larger than Song Thrushes. The head and upperparts are greyish-brown in colour, with a slightly paler rump. The tail is dark with white tips when seen in flight. The face is pale grey-white, with a black border to the ear-coverts and a bold white eyering.

The wings show black feather centres with broad, pale grey fringes on the coverts, tertials and secondaries. The primary feathers are black with a thin pale fringe. The long tail is dark grey-brown, with white outer tail feathers.

This view of an adult Mistle Thrush shows just how 'pigeon-chested' they can look, particularly with the wings and tail held drooped, nearly to ground level. The bold spotting and cold grey-brown colour of the upperparts are immediately clear, as is the yellow wash on the flanks.

This view shows the bold black spots on the underparts, on a white to yellowish background colour. The undertail-coverts are white, with just a few, if any, marks. The bill is black with an orange base, and is stouter than that of other thrushes. The eye is big and black, and the legs are pink.

The large size, undulating flight and resounding 'football rattle' call of Mistle Thrushes are instant giveaways when seen in the air. They appear quite grey with a pale head. The buffy-grey rump contrasts strongly with the dark tail, and the white tips on the tail are easily seen.

Male Mistle Thrushes will always head to the top of a tree to sing. The song is similar to that of the Blackbird, but tends to be more raucous, quicker and a little lacking in melody. The male will begin singing in late February or early March, and the breeding season will last into June.

Juvenile Mistle Thrushes are basically similar to adult birds, but differences are easy to note. Youngsters appear quite spotted and streaked on the whole of the upperparts. The face pattern appears plainer than on adults and the underparts show a strong pale yellowish-buff wash. The wings are very well marked, even more so than those of the adults.

GARDEN WARBLER
(13-15CM, 5-6IN)

THE GARDEN WARBLER IS A RATHER PLAIN BIRD which is a common summer visitor to the whole of Europe, with the exception of parts of Ireland and the very southernmost areas of Europe. Not an easy bird to see, the Garden Warbler favours areas of scrub, particularly alongside woodland fringes or in dense hedgerows and undergrowth.

A skulking, rather shy bird, the Garden Warbler is usually best located by its pleasing soft song, very like the Blackcap's but a little quicker and longer.

Garden Warblers are rotund, plump birds, with a decidedly rounded head, short tail and short stubby bill. Their plumage can appear fairly nondescript, but a close view will reveal some more subtle tones. The large eye and plain face give the Garden Warbler a particularly gentle expression.

This species is closely related to the Blackcap and has similar habitat preferences. Blackcaps arrive earlier in spring than Garden Warblers and will chase the latter out of their territory. Both are very fond of eating blackberries in autumn.

The head of the Garden Warbler is pale brown, with an indistinct off-white supercilium and pale buff ear-coverts. The throat is off-white. The mantle is similar in tone to the head, but the rump is slightly paler. The wings are grey-brown with darker flight feathers. The tail appears dark. The underparts are off-white with a buff wash on the upper breast and flanks. The stout bill is dark grey, with a paler base. The large eye is black, while the legs and feet are greyish.

Garden Warblers can be very difficult to see, as they hide themselves deep inside a thicket or cover. Often a harsh 'tac-tac' can be the first clue to a Garden Warbler's whereabouts, and then the male will work his way through the undergrowth towards the top of the bush and begin to sing. It can be difficult to separate the similar songs of Garden Warbler and Blackcap, but, with practice and a good ear, the differences will become clear.

The juvenile Garden Warbler is basically the same as an adult bird, differing only in having 'fresher'-looking plumage, being browner on the upperparts and washed a stronger buff on the underparts. The bare parts are as those of the adult.

T HE BLACKCAP IS A STRIKING BIRD – long but quite slender-looking – with an easily recognisable plumage. Blackcaps are generally summer migrants breeding across the whole of Europe, but they are now also a regular and an increasingly common winter visitor, particularly in milder areas of Britain and the warm climes of southern Europe.

Blackcaps are quite hard to track down and locate once trees come into leaf, but they can usually be located by their rich, melodious song, full of strong phrases and snatches of mimicry.

The plumage of both male and female Blackcaps is instantly recognisable, and their slim lines, long body and song are distinctive whatever the quality of the view.

Female Blackcaps differ from males in having a rich russet-brown cap, a browner wash to the upperparts, browner wings and a greyish-brown wash to the underparts. The bare parts are similar to the male's, with dark silvery-grey legs and feet.

The male Blackcap is a striking bird and should not be confused with any other species. The head shows a glossy jet-black cap, contrasting with silvery cheeks and nape. The upperparts are entirely silvery-grey, becoming slightly darker towards the tail tip. The wings fade from grey on the shoulders to dark brown on the primaries. The underparts are whitish, with a greyish wash to the breast and flanks. The relatively long, thin bill is dark grey, the eye is big and black (with a white orbital ring) while the legs and feet are dark silvery-grey.

Blackcaps are not renowned for singing from the treetops, preferring to advertise themselves from the cover of a hedge or other tallish vegetation. The song is, as already mentioned, a melodious rich combination of warbles and mimicry, not unlike that of a Garden Warbler, which tends to have a longer song.

A young male Blackcap can be identified by its softer-coloured underparts, brownish-washed upperparts and, most significantly, a dark cap which shows mixed black and brown feathers.

CHIFFCHAFF
(10-11CM, 4-4½IN)

T HE CHIFFCHAFF IS A SMALL GREENISH WARBLER which is common across much of Europe in the summer and can now be found also, in Britain and parts of Europe, as a winter resident.

During the autumn, the British and western European population of Chiffchaffs is bolstered by Chiffchaffs from Scandinavia and even Siberia. These northern and eastern birds are generally greyer than the more olive-brown individuals seen regularly in Britain, but it takes an experienced eye to pick up these immigrants !

The Chiffchaff is almost identical to the Willow Warbler. Differences can be detected, however. Carefully look at the shape and proportions of the bird, study the subtleties of plumage detail and, if you are still struggling, listen for the song, a monotonous, repeated 'chiff-chiff' heard from March to June and less so in autumn. The contact call is a plaintive 'hueet'.

The Chiffchaff is *slightly* smaller than the Willow Warbler and structural points to look for are the more rounded head, shorter wings and more stocky appearance. The head and upperparts are dull olive-green, with a greyish-white supercilium, blackish eyestripe and pale-looking ear-coverts. The uppertail is dark olive-green. The wings are also olivey, but with darker feather centres, and very dark primaries. The underparts are buffy-white except for a white throat. The thin bill is dark with a faint pale base. The eye is black, with a thin white orbital ring, and the legs and feet are generally dark brown, unlike those of Willow Warbler.

Chiffchaffs are seen in almost any type of woodland during the breeding season, but during the winter the Chiffchaff is a frequent visitor to gardens in urban or rural settings. This summer-plumaged Chiffchaff shows how alert this species can be, often moving quickly through cover, constantly flicking its wings and jerking from branch to branch.

In the autumn, Chiffchaffs can moult into a brownish or olive guise. Through plumage wear, adult birds fade to brownish-grey on the upperparts and whitish on the underparts. After moult in July to September, they can look olivey once more on the upperparts and the underparts may show a very diffuse yellowy wash to the breast and flanks.

Young Chiffchaffs appear to be quite brightly marked birds, rich olivey-green on the upperparts and washed yellow below. After their first moult in the autumn, they soon begin to resemble a grey-washed adult.

 T HE WILLOW WARBLER IS A SMALL, GREENISH WARBLER which is very common right the way across Europe in the summer months, arriving in late March or early April and departing in September and October. Unlike those of its close relative the Chiffchaff, Willow Warbler wintering records, in Britain at least, are still exceptional.

 Willow Warblers are found in the same habitats as the Chiffchaff, but tend to favour smaller, younger trees, bushes and ground vegetation. They can also be seen in quiet woodland-fringe gardens, or in those which are somewhat overgrown.

 A small bird, the Willow Warbler bears a very strong resemblance to the stockier Chiffchaff. Check the structural points – look at the long wing length of the Willow Warbler (short on Chiffchaff) – the leg colour (pale on Willow Warbler, dark on Chiffchaff) and overall plumage tones. If all else fails, listen for the song, which is a series of fluid descending notes, ending with a rapid flourish. The call note is similar to that of the Chiffchaff, but is a more penetrating and musical 'hoo-eet'.

Willow Warblers in spring are generally paler in appearance than Chiffchaffs. The head and upperparts are pale olive-green, with a yellow-tinged long supercilium, not short and buff as on the Chiffchaff. A pencil-thin black eyestripe and slightly blotched cheeks are also apparent. The wings are pale olive, except for darker primaries. The tail is also dark. The underparts are cleaner than the Chiffchaff's, lacking any buff tones. The breast and flanks are washed pale yellow and the belly is whitish.

The bill shows a dark upper mandible and tip, with a flesh-pink lower mandible. The eye is black, with a less obvious ring than the Chiffchaff and the legs are pale flesh.

Occasionally, in Britain, you may encounter a very pale-looking 'northern' Willow Warbler. The whole of the head and upperparts are a washed-out pale greenish colour, except for darker primaries, tail tip and eyestripe. The supercilium and underparts are off-white, with a greyish suffusion on the flanks. These striking birds may invite confusion with rarer warblers, so beware!

Basically resembling a spring bird, the adult Willow Warbler in autumn may look a little browner on the upperparts and more yellowy below.

Some young Willow Warblers in autumn can be particularly striking. The upperparts are pale olive-green, while the underparts are rich lemon, from the face and throat to the undertail-coverts, although the belly usually stays whitish. The bare parts are much as the adult's.

SPOTTED FLYCATCHER
(13-14CM, 5½-6IN)

T HE SPOTTED FLYCATCHER IS A LONG-WINGED, ACTIVE, SUMMER MIGRANT which is found throughout Europe between May and October. Favourite habitats for this perky bird include suburban gardens, parkland and open woodlands.

Spotted Flycatchers are often seen darting out from cover in search of food, but when not sallying to and fro the Spotted Flycatcher will spend long periods of time on an exposed twig or branch, sitting quietly whilst pumping its tail before SNAP – a suitable meal passes by.

The Spotted Flycatcher is a sparrow-sized bird which, like all other species of flycatchers, has long wings, a squarish tail, short legs, big eyes and a broad-based bill. The plumage is somewhat drab, but the combination of grey, brown, and white streaks and characteristic behavioural traits means that the Spotted Flycatcher can be readily identified.

The adult Spotted Flycatcher's face shows an indistinct white supercilium, with greyish cheeks and a white moustache. The throat is off-white, with thin dark streaks which extend onto the breast. The rest of the underparts are white. The bill, eye and legs are black. The call is a thin 'tzee' and the, frankly, poor song is made up of several wheezy notes.

The crown is pale grey with dark streaks, merging into the greyish-brown nape, mantle and rump. The tail is dark, with white outer tail feathers. The wings show pale edgings to all the main feather tracts. The tertials, secondaries and primaries are all dark brown.

In flight, Spotted Flycatchers are very agile and acrobatic as they catch assorted flying insects, including bees, butterflies and greenfly. They often twist and turn, close to the ground, before returning to the same perch.

After a moult in the late summer, the first-winter Spotted Flycatcher basically resembles an adult bird, except for browner upperparts, broader edges to the wing and a prominent bar on the greater coverts. The underparts are much as those of the adult, as are the bare parts.

Juvenile Spotted Flycatchers have a heavily scalloped head, mantle and breast, contrasting with the broadly fringed grey-brown wings and dark tail. The underparts are white. The bare parts are as the adult's.

GOLDCREST
(8-9CM, 3½IN)

T HE GOLDCREST IS A TINY ROUND BALL OF FEATHERS, and is the smallest bird to be found in Europe. Goldcrests are widespread across Britain and much of Europe with the exception of the far northern reaches of Scandinavia. They can be seen in many gardens, as well as hedgerows, bushes and woodland, especially coniferous forests.

Goldcrests are agile little birds, flicking constantly from bough to bough, picking at aphids or flycatching. When not moving through the treetops, Goldcrests are equally at home 'foraging' through small bushes and grass.

Aside from the obvious tiny size, Goldcrests are strikingly marked birds and the males and females, when seen well, can be easily separated. In the winter months, Goldcrests will often join up with the local roving tit flock, as they search for food through gardens and woods.

The male's song is a distinctive, flourishing affair, a 'seeh, zeeda-zeeda-sissisyn-see'. The call is a high-pitched, short, rapid 'zee zee zee'.

The head of the Goldcrest shows a pale greyish face, which contrasts with the olivey cheeks, nape and sides of the crown. The centre of the crown shows an obvious orangey-yellow stripe, bordered black. The mantle and rump are olive-green, with a slightly darker tail. The wings are blackish, with bold creamy wing bars and edges to the feathers. The underparts are buffy-white, washed grey. The tiny, thin bill is black, the large 'surprised'-looking eye is also black, while the legs and feet are dark orangey brown.

The female Goldcrest basically resembles the male except for the crown stripe, which is bright yellow, lacking any of the male's orange tones.

When displaying, the male Goldcrest will raise his crown feathers as a signal to the female. The 'rippling' effect is very striking as the crown feathers are spread to reveal the orange feathering.

Ever active, the Goldcrest flicks its wings continuously during a seemingly relentless search for food among trees. They are particularly partial to flies and spiders and will also eat greenfly, beetle larvae and moths.

The juvenile Goldcrest is a dull version of the adult. The upperparts are browner-tinged and they lack any sort of markings on the crown, except for perhaps some black on the crown sides. The bare parts are as the adult's.

LONG-TAILED TIT
(12-14CM, 5-6IN)

T HE LONG-TAILED TIT IS A DELICIOUS LITTLE BIRD which is found commonly across Europe. As well as wooded gardens, Long-tailed Tits tend to favour woodland fringe, scrub and dense hedgerows.

The appearance of Long-tailed Tits varies somewhat across Europe, from white-headed birds in Scandinavia to black-streaked birds in Spain. Long-tailed Tits are busy birds, always on the move, on a seemingly constant search for food. Family groups join together in the autumn and winter, and the large feeding parties trill their way through gardens and woods, often with other species in tow.

The tiny oval body, rounded head, short stubby bill, distinctive plumage pattern and massively long tail make the Long-tailed Tit an unmistakable delight. The constant contact call, a piercing 'tsee, tsee, tsee' is easily recognisable once learnt.

Long-tailed Tits do not eat as many seeds as other tits, preferring to eat insects. Highly agile in their search for food amongst the leaves and twigs, they hang upside-down by one foot while clutching food in the other. They have an incredibly small black bill and black eye with a red orbital ring. The small, rarely seen feet and legs are dark grey-brown.

Long-tailed Tits in Britain and Ireland have an off-white head with two broad blackish bands running from the bill base, over the crown to the nape. The mantle and rump are blackish. The long tail is black with white outermost feathers. The wings show pink scapulars, broad white edges to the tertials and pale-tipped dark flight feathers. The underparts are off-white, with a pink flush to the flanks and belly. The undertail is spotted white on black.

▶ A familiar sight after young have fledged is to see and hear a flock of Long-tailed Tits streaming through a woodland. They are unmistakable, with tiny bodies and long tails.

▶ The small, perfectly round nest of the Long-tailed Tit is a marvel to behold. Made of lichen, moss, feathers and anything soft, the birds build the nest deep inside cover, particularly prickly bushes, or in a tree fork.

▶ In Scandinavia, Long-tailed Tits have unmarked snowy-white heads, paler pink scapulars and more white in the wings.

◀ Birds on the near European continent show broader black head stripes than those of British birds and the flanks are a dingier, dirtier pink. Further south and into Spain, Long-tailed Tits become darker and dingier still.

▶ Juvenile Long-tailed Tits are browner than the adults, with more dark markings on the head, whitish (not pink) scapulars and a shorter tail. The underparts are white, with little or no trace of pink. The bill shows a yellowish base, but is otherwise black. The eye may appear brown, but the red ring is as the adult's, as is the leg colour.

T HE MARSH TIT IS A SMALL COMPACT BIRD, common across much of western Europe. It is, however, absent from Ireland, and most of Scotland and Scandinavia. The Marsh Tit can be seen in woods, especially damp broadleaf areas, copses, parks and gardens.

Marsh Tits are less inclined to move in large groups, although they do join roving winter flocks, and are more than happy to come to garden bird feeders.

Although generally a thickset-looking bird, the Marsh Tit can look surprisingly sleek. Notice the thick neck, the stubby bill and round head and then exercise extreme caution! Pay close attention, using the points listed below, to ensure that you are not looking at the near-identical Willow Tit. Listen for the all-important call. Marsh Tit's is a very obvious nasal 'pitchou' or 'pitchou ke ke ke'. The song, seldom heard, is a typical tit-like 'chip chip' – rapid and ringing.

When seen from the back, the thickset neck and smallish head are apparent. Also, the wings look plain, lacking any sort of wing panel, which Willow Tit shows. As the bird's plumage wears in late summer, the brown gets darker.

Marsh Tits have a neat, glossy-looking cap, which extends to the rear nape. The cheeks are white, fading to buffy behind the ear-coverts. Notice the small black bib on the chin. The upperparts are wholly warm brown, except for the darker square-ended tail. The wings are also brown, with darker primaries. The underparts fade from off-white on the upper breast to buffy on the lower breast and flanks. The stubby bill and the large eye are black, while the legs are slightly greyer in tone.

Marsh Tits are frequent users of nest holes, but unlike the Willow Tit they do not excavate their own holes, preferring to make use of natural cavities for nesting. Marsh Tits will nest between April and June, producing one brood of six to eight young. They will also nest in rotten stumps, holes in walls or even holes in the ground.

In Scandinavia, paler Marsh Tits are found, exhibiting wing panels, a feature associated with Willow Tits. However, all the structural points are the same, as is the voice. Exercise caution on these!

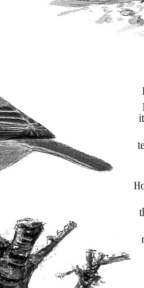

In the early months of the year, the plumage of the Marsh Tit will be at its most pristine, and on occasion it can resemble the Willow Tit. The tertials and secondaries on the wing of the Marsh Tit may show subtle pale edges, as on the Willow Tit. However, the markings on the Marsh Tit should always be less obvious than on a 'spring-plumaged' Willow Tit, and there are the previously mentioned plumage, structural and vocal differences to check and double-check.

 HE WILLOW TIT IS A SMALL COMPACT BIRD which, like its very close relative the Marsh Tit, is common across much of western Europe, with the exception of Ireland and much of Scotland. Unlike the Marsh Tit, this species is also widely spread across much of Scandinavia. The name is inappropriate since the Willow Tit is not exclusively 'tied' to willows; favourite habitats include damp areas of woodland, especially alder and birch scrub, and also conifer areas. Willow Tits can also be seen in hedges and occasionally at garden feeders.

When confronted with a possible Willow Tit, remember to exercise extreme caution. If you think the bird before you is a Willow Tit, check all the relevant structural and plumage characters carefully. Marsh and Willow Tits are very similar indeed.

In Britain, the flanks of Willow Tits are washed buff, markedly stronger than on Marsh Tits. The short and stubby bill is black, as is the eye. The legs and feet are dark silvery-grey.

Willow Tits are quite rounded birds, having a particularly thickset appearance, with a large head and 'bullnecked' look, which with practice can become quite a distinctive feature. As well as structural differences, pay close attention to the face and wing markings, and to the call, which is a loud, deep, buzzy 'tchay-tchay-tchay', while the song is a slow, sad, 'tsui, tsui, tsui'.

The Willow Tit has a dull black cap that extends to the rear nape and mantle. The cheeks are *wholly* white (not white and buff as on Marsh Tit) and the bib on the chin is larger. The upperparts are the same colour as Marsh Tit's, except for the tail, which has pale outer feathers. Notice also the shape of the tail of Willow Tit – it is slightly rounded, as opposed to square-ended on Marsh Tit. Fresh-plumaged Willow Tits show pale edgings to the secondaries and tertials, forming a distinctive panel, lacking on Marsh Tits.

▶ Willow Tits nest in holes which they tend to excavate themselves in rotten tree stumps, unlike the 'lazy' Marsh Tit, which uses natural holes. Problems arise, however, when Marsh Tits take over a Willow Tit nest hole!

▲ When seen head on, the Willow Tit, although proving tough to identify, does show some of those all-important features needed to separate it from Marsh Tit. The large-headed look may be apparent, as will the dull crown. Concentrate on the size of the black moustache, as it will always be larger and broader on the Willow Tit.

▶ In Scandinavia, the Willow Tit is a paler bird than that found in the rest of Europe. The mantle and wings are greyer, which accentuates the paleness of the wing panel. The tail also appears greyer, and the white outer tail feathers are readily apparent. The cheeks and underparts are white.

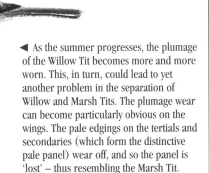

◀ As the summer progresses, the plumage of the Willow Tit becomes more and more worn. This, in turn, could lead to yet another problem in the separation of Willow and Marsh Tits. The plumage wear can become particularly obvious on the wings. The pale edgings on the tertials and secondaries (which form the distinctive pale panel) wear off, and so the panel is 'lost' – thus resembling the Marsh Tit.

COAL TIT
(10.5-11.5CM, 4½IN)

THE COAL TIT IS A CHARACTERISTIC SMALL-BODIED, BUT LARGE-HEADED TIT which is widespread across the whole of Europe with the exception of northernmost Scandinavia. Commonly seen visiting gardens, where it is very partial to nutbags, the Coal Tit is found in coniferous woodland in considerable numbers, and also in deciduous woods and hedges.

The smallest tit found in the UK, the Coal Tit is a very acrobatic little bird, always moving and appearing restless. Although they will join up with other tits in the winter months, Coal Tits tend to be a little more solitary, mixing mainly with other Coal Tits.

A curiously proportioned bird, the combination of large head, short forked tail and thin bill is distinctive. The plumage, despite 'racial variants', is constant in pattern (if not colour) and should not be confusable. The song is a very distinctive, far-carrying 'pitchou pitchou pitchou' or 'tisui-tisui-tisui' and the call is usually a thin 'tseu'.

The agility of Coal Tits is something to marvel at. Any angle is possible, and this view illustrates not only the dexterity of the Coal Tit, but also the head and back patterning splendidly. As well as being inveterate treetop feeders, Coal Tits are also capable ground foragers. They will probe all the ground in search of insects, or rummage through leaf litter for seeds.

The adult's large head shows a glossy black crown and large black bib, contrasting with bold white cheeks and rear nape patch. The mantle and rump are olive grey, like the tail, although the tip is darker. The wings are also olive grey except for two prominent white wing bars and tips to the tertials. Below the bib is a small white patch on the upper breast, fading into a pinky-buff flush on the remaining underparts. The bill, eye and legs are black.

Birds on the Continent, particularly in northern Europe, differ from British Coal Tits in having whiter cheeks, a greyer back and paler buff-pink underparts.

In Ireland, the adult Coal Tits show a yellowy wash to the nape and cheeks, and the upperparts are buffier in tone, while the underparts are flushed yellowy-buff.

Juvenile Coal Tits are drab affairs compared with their parents. The basic patterns are the same, but the colours, on the upperparts at least, are subdued. The mantle is washed brownish, with darker feather centres to the wings. The crown and bib are sooty or brownish. The white on the adult – cheeks, nape patch, wing bars and underparts – is all washed yellow, sometimes quite strongly. The rear end tends to be whiter. The bare parts are as those of the adult.

BLUE TIT
(11-12CM, 4½-5IN)

THE BLUE TIT IS A CHARACTERISTIC SMALL, ACTIVE LITTLE TIT which is commonly found throughout Europe. It has adapted well to man's influence, and is seen in every possible context, from the most remote country area to the busiest of cities.

A busy bird, the Blue Tit is one of the 'jolliest' visitors you will see in your garden, providing endless entertainment as it searches out food or nesting sites.

Blue Tits have a small pointed bill, useful for picking off insects and grubs, and they are highly acrobatic, often seen hanging upside-down on branches in search of another prey item. Their song is a charming, ringing 'tsee-tsee-tsirr' and their call a cheeky, but harsher 'churr-urr-urr'.

The adult Blue Tit has a sky-blue crown, contrasting strongly with the white and black of the rest of the head. A neat black line runs from the bill base, through the eye onto the rear crown, where it widens and falls down, before extending forwards to meet a neat black bib on the chin. The mantle and rump are lime-green, and the tail, slightly notched at rest, is dull blue-grey. The wings are sky-blue with a thin white wing bar. The primary tips are dark blue-black. The underparts are lemon-yellow, with a black central streak on the belly, which can be a very variable feature. The undertail is washed whitish. The bill and eye are black, as is the eye, while the legs and feet are bluish lead-grey.

▶ One of the best ways to enjoy Blue Tits is to attract them to your garden by way of a nestbox, where they may produce two or three large broods of youngsters in a season. This adult is removing a nestling's faecal sac to ensure that the nestbox stays clean.

▲ Juveniles are basically yellow, green and black, duller than adult birds, and are often found sitting in small groups. The head shows a blackish crown and green hindneck and the white of the adult is replaced by yellow. The mantle is bright green, while the wings and tail look more green than blue. The underparts are wholly yellow.

▼ In the autumn and winter, Blue Tits take on a less obvious role, joining up with other tits and woodland species as they rove through gardens, parks, woodland and hedgerows. If they struggle to find insects or seeds, they are happy to feed on fruit, such as this windfall apple.

◀ Perhaps *the* classic view of the Blue Tit is with its head dipping into the cream at the top of a milk bottle. The birds can spend several minutes pecking at the bottle top, picking small pieces off and tossing them to one side. Eventually – BREAKTHROUGH – and the milk is all theirs.

GREAT TIT
(13.5-14.5CM, 5½IN)

THE GREAT TIT IS AN INSTANTLY FAMILIAR GARDEN BIRD which is common throughout Europe. A real lover of the garden habitat, it is one of the most regular visitors to almost all birdtables, feeders, nutbags and bird boxes. Away from gardens, the Great Tit can be seen in almost any woodland habitat, parks, hedgerows and even reedbeds.

The largest member of the family, the Great Tit therefore lacks some of the agility of its smaller relatives. It is, however, a great vocalist, imitating a wide range of different species. Constant calls can also be heard and these include a loud resonant 'teecha-teecha-teecha', which represents the Great Tit's song, and a ringing 'zinc, zinc'.

Female Great Tits are noticeably duller than males. The mantle tends to be paler green, the wings greyer, with greener fringes, while the black on the head and throat is duller. The underparts are still yellow, but the central strip is narrower than on the male. When seen clearly, the differences in underpart patterning are easily noted. The female's thinner black band is sometimes broken or flecked with white or yellow.

Males show a glossy black head, with a bold white cheek patch and pale yellowish nape. The mantle is olive green fading to powder blue on the rump. The tail is brighter blue, with prominent white outer tail feathers. The wings are blue grey with a white wingbar and feather fringing. The belly is bright yellow with a broad, black central band. The vent and undertail are white, with a thin black central stripe. The bill and eye are black, while the legs and feet are blue grey.

130

▼ The male Great Tit starts his characteristic display routine as early as February. With his familiar song he tries to entice a suitable mate and the pair will, hopefully, raise a brood or two during the breeding season.

▶ In the winter the Great Tit becomes a more greedy feeder. After a summer of feeding youngsters and themselves on a diet of caterpillars, larvae and insects, they relish the winter months, devouring all sorts of nuts, fruit or seeds that they can find. Despite their comparatively large size, they are still quite agile feeders, particularly on nutbags.

▶ Note how this female, in characteristic pose, has a far narrower band on the underparts than the male (above) but it is distinctly stronger and blacker than that of the juvenile (below).

◀ Juveniles are washed-out affairs, much duller than either adult. The black on the head is replaced by sooty-brown, the cheeks are washed yellow, the nape patch is off-white, while the upperparts appear very dull olive-brown. The wings are duller and more olive, with buffy-white fringes. The softer yellow underparts show a sooty-brown central stripe. The bill shows a yellowy gape, otherwise the bare parts are the same as those of the adults.

131

NUTHATCH
(13.5-14.5CM, 5¹/₂-6IN)

T HE NUTHATCH IS A CHARACTERISTIC, SLEEK, WOODPECKER-LIKE BIRD which has a somewhat irregular distribution across Europe. In Britain the Nuthatch is a common bird south of a line running from the Solway to the Tees, but is absent from Ireland. It is also missing from much of Scandinavia and areas of southern Europe. The Nuthatch is seen in parkland and deciduous woodlands with mature trees, as well as in gardens.

An extremely active bird, the Nuthatch shows remarkable agility as it climbs along branches and trunks in every possible direction – up, down and around. However, the Nuthatch is not solely a tree feeder, and is also seen feeding on the ground with distinctive jerky hops. It frequently visits garden birdtables and nutbags and is not a particularly shy bird.

A neat bird, the Nuthatch has distinctive markings as well as a compact torpedo-shaped body, short tail and hefty sharp bill. They have entirely steel blue-grey upperparts and wings, with the exception of the leading edge and flight feathers of the wing, and the tail shows black and white sides. A broad black eyestripe extends from the bill base to the ear-coverts, contrasting with the white throat. The underparts are buffy (but vary on Continental birds), with the male showing chestnut flanks, paler on the female. The bill is blackish with a silvery base, the eye is black, and the legs and feet are pinkish. Young birds have brown-tinged upperparts, a narrower, duller mask and dull brown flanks.

The Nuthatch's flight is undulating and, along with the size and shape, it gives an initial impression of a small woodpecker. However, a good view reveals the distinctive grey upperparts and orange-buff underparts.

The Nuthatch uses its sturdy bill to great effect when storing or retrieving food. Several hefty blows will manage to break open a stored nut wedged in a bark crevice, and the tapping can be heard for several hundred metres in a quiet woodland.

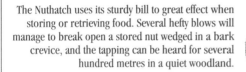

Nuthatches are quite capable of foraging through autumn leaf litter when searching for nuts, seeds or insects. When frightened they will take flight, head for a safe bough and call a loud, full-sounding and excitable 'chewit chewit'.

TREECREEPER
(12-13CM, 4½-5IN)

THE TREECREEPER IS A DISTINCTIVE TINY MOUSE-LIKE BIRD which is common in Britain, north and eastern Europe. It is found in woodlands, parks and gardens.

A shy and unobtrusive bird, the Treecreeper moves from tree to tree with an undulating flight, often starting at the base of a tree, moving stiffly upwards and around, with the tail held firmly against the trunk and feet well spread.

Treecreepers have a thin curved bill and a stiff pointed tail. The bill, quite short in reality, shows a dark upper mandible and a pinky lower mandible. The large eye is black and the legs and feet are pink.

The call of the Treecreeper is a high-pitched, thin 'tsee' which can be quite tricky to pick up, particularly in a bird-filled wood.

Treecreepers are streaked finely with white from the head to the mantle. The ground colour of the upperparts is dark brown, and they can be very well camouflaged against a tree trunk. A longish, white supercilium is fairly prominent, and look out for a dark ear-covert patch with cream centre. The rump is rufous-brown, fading to dark brown on the tail tip. Notice the distinctive tail shape and the way the Treecreeper uses its tail to 'clamp' onto the trunk. The wings show a very complex pattern – black, creams, buffy-yellow and grey-browns, – and can look very striking, particularly in flight. The underparts are silky white, with a faint buff wash to the flanks.

The upperparts of a Treecreeper in flight show the rump contrasting strongly with the brown mantle, while on the wing a broad creamy-yellow band can be seen across the middle. The curved bill is also obvious. Notice the deeply notched tail.

The Treecreeper can be a particularly elusive species in the breeding season. It nests amongst big clumps of ivy, behind loose bits of bark or in the cracks of trees. Here is an adult bird returning to the nest after a feeding trip, about to pass on its goodies to the emerging juvenile. The adult bird would have worked a familiar pattern up and around the trunk and branches of one tree, picking insects from the bark crevices, before dropping to the bottom of another and working through the whole procedure again.

Juvenile Treecreepers are very similar to the adult birds, but can, with care, be told apart. The upperparts lack some of the warm brown tones of the adult's, showing a greyish-looking base colour, with fine streaks on the head. The mantle often looks quite spotted and buff, unlike the adult's. The rump looks far less reddy-brown. The underparts are generally white, occasionally washed buff. The breast and flanks may show some faint brown flecks. Soon after fledging the youngsters will be scurrying around trunks and branches, just like their parents.

JAY
(33-36CM, 13-14IN)

THE JAY IS WITHOUT DOUBT the most colourful and most spectacular member of the crow family to be found in northern Europe. The bird has a widespread distribution throughout the region, although it is absent from the very north of Scotland and northern Scandinavia.

The chunky build, stout, heavy bill and colourful plumage make this generally shy bird unmistakable. Jays can be seen in gardens on the edges of woodland with some regularity, and they can also be found in local parks and plantations. In exceptional years, Jays from the Continent flood across the North Sea or English Channel in the autumn en masse into Britain. It is thought that food shortages force the birds to 'erupt' in such a spectacular manner.

As Jays are such wary birds, it is not surprising to see them sitting nervously at the edge of woods, awaiting their chance to drop into a garden for food.

The head shows a whitish forehead, speckled with fine black streaks, fading on the top of the crown into a darkish fleshy-pink nape and cheek. A fat black moustache contrasts with the gleaming white throat patch. The mantle and upper half of the closed wing are brownish-pink. The bird's 'elbow' is a beautiful aquamarine, notched subtly with black and white, and this contrasts with the black and white remainder of the wing.

The rump is snowy-white and the longish tail is black. The underparts are a delicate pink, fading to white on the undertail. The eye is orangey-brown, while the stout bill is blackish, with a paler base. The legs and feet are fleshy-pink. The bird at the rear is particularly alert, with crown feathers raised, revealing the black streaking.

▼ As with other members of the 'Corvid' group, Jays can make themselves very unpopular with the general public because of their habit of pilfering nests for young or sitting birds. Here the victim is a fairly defenceless Wren, who in this situation really is not going to fare too well.

▲ When seen from underneath, the Jay does not display its spectacular plumage. It flies with a slow, quite deliberate undulating flight action. However, the longer the flight, the more unsteady their course becomes and the more laboured the action appears.

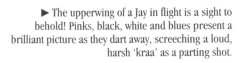

► The upperwing of a Jay in flight is a sight to behold! Pinks, black, white and blues present a brilliant picture as they dart away, screeching a loud, harsh 'kraa' as a parting shot.

▼ During the autumn, Jays often congregate in small groups in our woodlands to collect and store nuts for the winter. They hop clumsily over the ground in search of acorns, which, when in excess, they bury.

MAGPIE
(42-50CM, 16½-20IN)

THE MAGPIE IS A VERY DISTINCTIVE MEMBER OF THE CROW FAMILY and is common and widespread throughout northern Europe. It can be found in a wide variety of habitats – from hedgerows to coastal bushes, moorland to woodland, urban or rural settings.

Noisy chattering birds, Magpies have a reputation, well founded it must be said, as a garden bird terrorist. No garden nesting species is safe from this pied robber, but the result is that the other birds have learnt to conceal their nests more effectively from Magpies.

The black, white and iridescent plumage, chunky, stout black bill and long tail make Magpies unmistakable. A male Magpie's tail is generally longer than the female's, and some think that the length of tail determines 'rank' in a group.

When on the ground, Magpies hop and sidle along, looking decidedly shifty! The tail is always held up when they walk.

Magpies show an entirely jet-black head, back, rump and breast. The belly is bright white, contrasting with the black undertail. The wings show a bold white scapular patch, with the remainder of the wing being very dark, but with a delicate blue-green iridescence. The long, long tail has a gleaming green sheen, bluer towards the tip, but at a distance the iridescence of wings and tail is not apparent, and they appear black. The eye is black and the legs and feet are pale lead-grey.

◄ In winter it is not uncommon to see flocks of dozens of Magpies leaping through treetops, constantly chattering raucously and always appearing as though they are looking for trouble!

► Magpies fly with a combination of quick beats interspersed by glides. The white of the flight feathers and the prominent white 'V' on the back contrast strongly with the rest of the dark plumage. This bird shows a pale rump, a feature that is variable on Magpies, from whitish to blackish-brown.

▲ Baby Magpies, just out of the nest, look a picture of innocence, sitting unobtrusively on top of a hedgerow. Very short-tailed, the youngsters can stay calm and collected for long periods of time, whilst awaiting the adults' arrival with food.

◄ A juvenile Magpie is basically identical to the adult, except for a shorter tail, less glossy wings and tail and sooty-looking head and breast. The white scapulars and belly are duller, but after their first moult in August or September, the gloss begins to show.

JACKDAW
(32-34CM, 12½-13½IN)

T HE JACKDAW IS A SHORT-BILLED COMPACT MEMBER OF THE CROW FAMILY, commonly found across most of northern Europe in a wide variety of habitats, from the countryside to the city.

Extremely inquisitive birds, Jackdaws have an air of self-confidence as they strut boldly on the ground or over roofs, and they prove themselves to be adept fliers, tumbling, sometimes *en masse,* out of the sky, before gliding great distances on suitable air currents.

Despite initial appearances of being 'just another black crow', Jackdaws are, in fact, stylish-looking birds. The basic points to look for on the various members of the crow family are size, bill shape and colour, and plumage detail. Also, in flight, notice the wing and tail shape, and whenever possible listen for the call.

Jackdaws often announce their presence with a resonant, quite high-pitched 'keya' or a harsh 'chak'. Adult Jackdaws, like other crows, mate for life.

▶ One of the most familiar views of a Jackdaw is to see them perched on top of a chimney stack. The head shows a glossy black cap and bib, contrasting strongly with the pale grey nape. The upperparts are sooty-black, with, in certain lights, a strong purple-green gloss, particularly on the wings. The underparts are dark grey, particularly on the breast.

▼ The bill, quite short when compared with other crows, is silvery-black, as are the legs and feet. The eye is a piercing silvery-white colour. Continental Jackdaws appear less grey on the nape and breast, while birds from northeast Europe are grey on the nape but with a whitish patch at the base of the neck.

▲ Jackdaw roosts look spectacular. Birds can drop like a stone, out of the flock, and more will follow in a remarkably agile aerial display. In flight, Jackdaws keep their tail closed, but when soaring the tail is fanned and appears rounded.

▶ Jackdaws often breed in loose colonies in urban and rural settings. Urban birds nest in buildings such as churches or in parkland, while birds with a more rural outlook favour trees with suitable nest holes, as the pair illustrated here has done.

▼ A juvenile Jackdaw shows the same plumage patterning as an adult, but the colours and tones are duller, with the nape less contrasting.

T HE ROOK IS A LARGE, FAMILIAR MEMBER OF THE CROW FAMILY which is commonly found across the whole of Britain and Ireland, parts of western Europe and eastwards into Russia. It is a wintering species in southern Europe and is absent from all but the southernmost parts of Scandinavia. Rooks are found mainly in lowland areas, particularly in agricultural parts, with short-cropped grassy fields being a real favourite. Rooks can also be seen in city parks and gardens.

A very sociable bird, Rooks will often join up with flocks of Carrion Crows and, especially, Jackdaws. When they are on the ground, Rooks appear rather ungainly as they hop and waddle around, in search of seeds, insects and worms.

Rooks have an unmistakable appearance. Pointed bill, pale face, a very high forehead and, at the other end, a very shaggy 'trousered' look about the legs. They are somewhat more bulky in build than Carrion Crows, with a less 'tidy' look about them.

Adult Rooks are glossy black all over with a decidedly purple sheen, particularly on the head and wings. The base of the bill, the chin and the lores show as a large area of whitish bare skin, which can be seen at considerable range. The bill, slender and pointed, is silvery-black, the eye is black and the legs and feet are silvery-black.

◄ Perhaps the most 'familiar' view of Rooks is this one, of groups perched high amongst the bare trees of winter, congregating around the old nests as nightfall approaches. All around, the distinctive, gentle-sounding 'kaah' can be heard. Rookeries can hold many birds and there is a whole host of folklore attached to them.

► When seen closely, the flight shape of the Rook is very distinctive. The head looks long, the wings appear pointed and narrower than the Carrion Crow's and the tail is rounded or wedge-shaped.

▼ In farmland, Rooks are often seen in fields, searching out various food items, such as worms or root crops and potatoes.

► Young Rooks can superficially resemble Carrion Crows, but can be told by size, shape of the head and those trousers! The plumage is entirely dark brown, lacking the gloss of the adults, and the face is fully feathered. As the young Rook grows, the plumage begins to blacken except for the wings, which stay brown.

143

CARRION CROW
(45-49CM, 18-19IN)

T HE CARRION CROW IS AN ALL-BLACK, MEDIUM-SIZED CROW which is common across Britain, except for northern Scotland. It is also common throughout much of western and southern Europe. Although 'absent' from Ireland, northern Scotland and Scandinavia, the Carrion Crow is in fact present in these areas, in the guise of the distinctive 'subspecies', the Hooded Crow. The range of the Hooded Crow extends from Ireland, through northern Scandinavia to Russia and beyond. To complicate things further, hybrids are also found in parts of the range where the two forms overlap.

Both forms favour open areas of farmland, woodlands, hills, cliffs and moorland, but they are equally at home in city parks and gardens.

Carrion Crows have a heavy black bill, a round forehead and flattish crown, and a square tail and are generally fairly compact-looking birds compared with the larger Rook. Hooded Crows are structurally identical, but are instantly recognised by their 'pseudo' pied plumage.

Adult Hooded Crows show black on the head, breast, wings and tail. The rest of the plumage is battleship-grey, which when seen in flight can present a very striking image. On a close view, you may be able to see the very fine black streaks on the underwing, upper breast, mantle and flanks.

The plumage of the Carrion Crow is entirely black, with a hint of purple gloss. The chance of confusion with young Rooks is high, but note the heftier, rounder bill of the Carrion Crow, 'shorts' rather than 'trousers' (feather coverage of the leg) and a squarer tail. Listen out for the call, a loud, hard 'kraa', the archetypal 'crow' call in fact.

► When seen in flight, the square-ended tail and the distinctive bulge along the rear edge of the wing will easily separate the Carrion Crow from other members of the crow family. Some birds, such as the one on the left, can show some silvery tones on the wings and tail. When soaring, the tail looks rounded.

▼ The calls of the various members of the crow family are another big clue to identification. The Jackdaw has a fairly high-pitched 'ke-ya' or a repeated, metallic 'chak'. Amongst its repertoire the Rook has a familiar long, drawn-out 'kaa-aa'. This Carrion Crow is caught in familiar pose, body and head held horizontally whilst giving its 'caw', more raucous than the Rook's, and lower-pitched.

◄ The Carrion Crow is a particularly opportunistic feeder, with a remarkably well-balanced diet! It is quite happy to feed on fruit and grain as well as insects and worms, small mammals, birds' eggs, nestlings and any bit of carrion it can find – such as this unfortunate Woodpigeon.

STARLING
(20.5-22.5CM, 8-9IN)

THE STARLING IS A FAMILIAR BIRD COMMON ACROSS WESTERN AND NORTHERN EUROPE. Starlings are widespread breeders throughout the continent, from northern Scandinavia to many Mediterranean countries and, of course, Britain and Ireland. Many hundreds of thousands of birds that spend the spring and summer in the north and east of Europe move westwards in the autumn and winter months to swell an already bulging population.

They are prolific breeders, making use of any possible hole, in a wall, roof or tree – they really are not too fussy. Particular favourite haunts are short-cropped areas of farmland, clifftops and gardens.

The Starling is a medium-sized passerine, slightly smaller than a Song Thrush, with a slender pointed bill, peaked head and a familiar quarrelsome manner. It is unmistakable.

At a quick glance, Starlings look wholly blackish, but close views show that the head and entire underparts have a distinctive purple iridescent gloss to the breast, fading into bottle-green on the flanks. The upperparts are also very dark, with fine flecks on the mantle, rump and wings. Male Starlings show fewer spots compared with the female. The bill is yellow and the legs and feet are pink.

In winter, adult Starlings change slightly from their glossy summer garb. Both sexes appear quite spotty, with buffy spots on the upperparts, whitish on the underparts, females tending to have bolder markings. The bill changes from yellow to mainly brown.

When they begin moulting to their more adult-like first-winter plumage, young Starlings can appear most odd in a 'half and half' transitional mode. By October most of the juveniles have undergone their transformation.

The juvenile Starling appears wholly buffy grey-brown, except for a whitish throat, greyish mottling on the underparts and gingery fringes to the wings. The bill is silvery-black, the eye is black and the legs and feet are horn-brown.

THIS IS PERHAPS *THE* MOST FAMILIAR, AND CERTAINLY THE COMMONEST, of all the birds you will see in your garden, wherever it is, be it in the most remote part of the countryside or in the deepest depths of the inner city. Its numbers, however, have recently shown a decline in parts of Britain.

The House Sparrow is frequently discarded by birdwatchers as something of an uninspiring bird, unattractive, unable to do anything other than 'cheep' and grub around, albeit cheekily, for crumbs. House Sparrows are, of course, far more interesting than we are commonly led to believe; colourful, full of character, accessible to all and full of field marks.

Although quite different in plumage, both sexes share a stubby bill (blackish in the male, paler in the female) and beady, black eye. The legs and feet are pinkish. The male House Sparrow is easy to identify, and should present no confusion with its close cousin the Tree Sparrow.

The distinctive feature of the male is the head, being a complex pattern of grey, black, brown and white. The crown and nape are grey, edged by rich brown which extends through the eye and nape. The lores, eye patch and chest patch are black. The cheeks are white, merging into an indistinct half-collar.

The female's head shows a broad creamy patch behind the eye, contrasting with a tawny crown, black eyestripe, greyish cheeks and white chin. The mantle is tawny-brown, paler on the back and rump. The tail is dark grey. The wings show straw-yellow 'tramlines', bordered by black, then merging into golden-browns, charcoal-blacks and white.

As part of his territorial display, a male House Sparrow can strike the most fetching of poses. The wings will droop, the tail will be cocked, the head held tilted and the black bib and chest markings will be fluffed up, resembling a magnificent beard! Squabbles are commonplace and House Sparrows are often seen tumbling in the dust.

Juvenile House Sparrows bear a close resemblance to the adult female, but are generally brighter owing to their fresh new feathers. The crown is darker, the supercilium richer honey-buff, while the markings on the mantle and wings all appear cleaner and richer in tone. The underparts are greyish-buff, and the bare parts are the same but for a yellowish bill.

In the winter, a male House Sparrow takes on a slightly different appearance. The grey crown is duller and flecked with white and pale grey, and the black bib becomes paler, again flecked white. The upperparts become duller and the bill changes colour from black to yellowish-brown.

TREE SPARROW
(13.5-14.5CM, 5½-6IN)

DESPITE BEING A MEMBER OF THE SOMETIMES MALIGNED SPARROW GROUP, the Tree Sparrow is in fact a particularly handsome bird. Widespread throughout Europe (although seemingly on the decline in Britain), Tree Sparrows can be seen in woodlands, farmland (arable land), as well as rural gardens.

Tree Sparrows are quite brightly coloured birds, and are a little more inclined to join finch and bunting flocks, outside the breeding season, than their commoner relative the House Sparrow. In flight, a resonant, quick 'tek-tek' call can be heard. With a clear view there should be little confusion between Tree and House Sparrows.

▼ In flight, the chestnut cap, black cheek spot, white collar and chestnut on the forewing should all be immediately apparent. With a prolonged view the yellowy pale brown rump is also noticeable, contrasting with the dark tail.

▼ When seen on the ground, the familiar sparrow shape is obvious – the big, round head, shortish wings and plump body – and the plumage is, surprisingly, quite bright. The sexes are similar. The head shows a bright chestnut cap, black lores, eyestripe and cheek patch, with a white neck collar. The mantle is a warm brown (sometimes tinged yellowish) with bold black streaking. The rump is unmarked buffy-yellow and the notched tail is dark brown to black. The wings show two white bars and a mix of rich chestnut and black. The underparts are white with light buffy flanks. The chunky bill is silvery-black, except for a yellow base, which can be more obvious on some birds than others, particularly in winter.

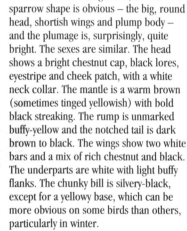

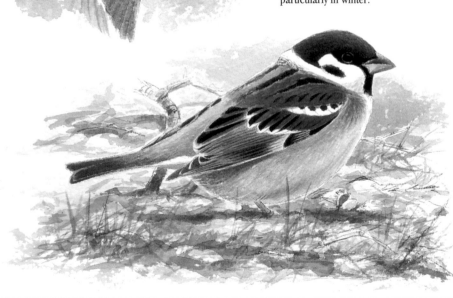

Young Tree Sparrows are simply a more subdued version of the adults. The head pattern is slightly different, with a greyish wash to the centre of the crown, merging into a less bright chestnut cap. The face pattern is similar, but is less bright, the cheek spot poorly defined. The upperparts are patterned the same, just duller on the young bird, and the underparts are greyish-brown. The black eye and pinkish feet and legs are just as on the adult, while the bill resembles that of a winter adult, blackish with a yellow base.

Tree Sparrows are far more likely to be encountered in rural settings, although they can be seen on the edges of villages and towns. Birds nest from April and through to early July, and raise two or three broods. Tree Sparrows nest in loose colonies, either in nestboxes or in excavated tree holes along woodland fringes.

CHAFFINCH
(14.5-16CM, 5½-6½IN)

 THE CHAFFINCH IS A STOCKY-LOOKING FINCH, widespread across Europe. It is a great user of birdtables and feeders, hopping around rather awkwardly on the ground in search of food.

Chaffinches breed throughout Britain and Ireland and right the way across Europe. Their range also extends into Scandinavia, and these birds migrate in the autumn to western Europe, where they join already vast populations.

The Chaffinch is a distinctively marked and shaped bird. The head shows a noticeable rear peak, while the body looks slightly pot-bellied and the tails is slightly forked. The blue-grey bill is comparatively thick and chunky and both sexes have a black eye and dull brownish-pink legs and feet. The male's bill is paler in winter, and the female's bill is always duller.

Male Chaffinches have a bluish-grey nape and crown which contrasts with a small black patch on the forehead and russet on the cheeks. The mantle is a dark reddish-brown, while the rump is a pale lime-green fading to a grey central tail. The rest of the tail is black with white outer feathers. The blackish wings show a prominent white shoulder patch and wing bar.

Females look far plainer than the male, lacking the bright colour tones. The head is greyish-brown, with a grey central stripe. The grey ear-coverts contrast with the fawn-brown face. The mantle is brownish and the rump patch is smaller and duller green than the male's, while the tail markings are the same. The wing markings are as on the male, sometimes a little duller. The underparts are greyish-brown, fading to whitish on the undertail and vent.

152

Chaffinches have a decidedly bounding and undulating flight, and any possible confusion with Brambling will be eliminated as soon as you see a flock of finches fly by. The double white wing bars and the greenish rump of the Chaffinch will be seen immediately, so precluding Brambling. Their call in flight is a soft 'chip'.

In winter Chaffinches become highly gregarious birds, joining to form substantial flocks roving around in search of food. Notice here the immature male (far right) showing some grey lines on the head and reddish lines on the mantle.

In the autumn and winter, male Chaffinches appear a fairly washed-out version of their summer selves. The blue of the head becomes grey, with a hint of brown, while the mantle and underpart coloration becomes noticeably faded. The black patch on the forehead is lost and the bill becomes whitish-grey, but retains the dark tip.

THE GENERAL SHAPE, ACTIONS, BEHAVIOUR AND 'LOOK' OF THE BRAMBLING are reminiscent of its close cousin the Chaffinch, but Bramblings show several plumage features which, given a reasonable view, make identification relatively easy.

Bramblings breed across Scandinavia, eastern Europe and into Russia, nesting mainly in birch trees or in conifers. In the winter months they move into western Europe in their thousands and feed, alongside other finches, on beechmast, seeds and berries. They can be found on farmland (around grain stores) and particularly in beech woodlands.

General plumage differences to look out for when separating Brambling from Chaffinch include a reduced amount of white on the wings and the tail, distinctive orangey-buff shoulders and breast (giving an all-over 'warmer' tone to the bird) and most obvious of all, particularly in flight, a narrow white rump.

The distinctive combination of colours sets the Brambling apart from all other finches. This winter-plumage male (male because of the black head, winter because of the greyish nape) shows all the characteristic features of the species – orange shoulders and breast, thin white wing bars and white rump patch. The mantle is scalloped black against a brown ground colour. The underparts fade from orange on the breast to white on the belly with spotted flanks.

Females, and some males in late winter, have a noticeably more drab head pattern than that of most males. The patterning is a mix of brown contrasting with grey on the side of the head and the nape. The mantle shows dark brown scalloping and the orange of the wings and breast is decidedly subdued.

154

◀ Winter flocks of Brambling are commonplace in some areas of Europe. Although a ground feeder, they can be quite nervous and will scare easily. As with this small group, they fly quickly up into the treetops before cautiously returning to the ground to feed.

▶ In flight the white rump patch is immediately apparent. Back-up features on a flying bird also include thin off-white wing bars and orange shoulders. This winter male shows the distinctive blackish head and back. The stubby bill and forked tail, common amongst finches, are apparent.

◀ A male Brambling in spring and summer will have a jet-black head and mantle, with none of the grey tones of a winter bird.

▶ Young birds lack even the greyish wash around the head of most females – brown and black are all they can muster. The orange on the shoulders and breast is still present, however, as is the white rump, so they are still readily identifiable.

BULLFINCH
(14-15CM, 5½-6IN)

THE BULLFINCH IS A FAIRLY COMMON, 'WELL-BUILT' BIRD which can be seen across Europe. Its favourite habitats include scrubby hedgerows, woodland fringe, orchards, plantations and, particularly in winter and spring, gardens.

Bullfinches have a scattered distribution across Europe. They are a resident breeding species (and rather sedentary) across almost all of western Europe. The birds in Scandinavia, especially those in the very northeast of the region, migrate in the autumn and winter in southward and westward directions, some of them occasionally reaching north Britain.

A fairly shy bird, the Bullfinch should present no identification problems when seen.

Bullfinches are seed-eaters, but they also take berries and are particularly fond of fruit-tree buds. Indeed, they can become a real problem to farmers of commercial fruit orchards.

Bullfinches are very round-looking birds, with a decidedly neckless look, a chunky deep-based bill, rounded wings and a square tail. The plumage is very striking, particularly the blend of colours on the male. Male Bullfinches of the British and Irish race show a glossy black cap, dark greyish mantle and scapulars, black wings with a broad white bar, a white rump and a square-ended black tail. The underparts, from face to belly, are carmine-red, fading to white on the vent and undertail. A most handsome bird.

Females share the males' black cap, but show a marked contrast between the grey nape and hindneck and the brownish-tinged dark grey mantle. The wings, rump and tail are as on the male. The underparts are pale fawn, washed lightly with pink. The vent and tail are as on the male. Both sexes have chunky black bill, black eyes and dark legs and feet. Here, this female is indulging in her favourite habit of nibbling fruit buds!

A very striking bird in flight, whatever the sex, the Bullfinch can be easily identified by its black cap, grey mantle of the male and brown mantle of the female, black and white wings, large white rump patch and black tail. The flight call is a soft, low, almost whistled 'decaw'.

Birds from the northern race differ from those in Britain in being larger, the males greyer above and lighter red below than the British birds, while northern females are also greyer on the upperparts and paler below.

Juvenile Bullfinches can be identified by their plain brown head and upperparts lacking the black cap and grey mantle, double off-white wing bars and pale brownish underparts. The bill also tends to be paler, more silvery-grey than black.

GREENFINCH
(14-15CM, 5½-6IN)

The Greenfinch is a stocky bird which is widely spread throughout northern Europe. It can be found in all manner of habitats, favouring more open areas, where it is very common, but Greenfinches are also to be seen in towns and cities, using gardens wherever they can.

Like all finches, the Greenfinch is an active bird, always on the move, but perhaps not quite as agile as its yellow and green relative, the Siskin.

Greenfinches have a heavy, stout bill and a short, deeply forked tail. The plumage is variable according to sex, from bright lime-green and yellow to brownish-green and yellow.

Both sexes have a beady, black eye, while that characteristic stout bill typical of a seed-eater is fleshy-pink, as are the feet and legs.

The male Greenfinch is generally green, with areas of yellow, black and grey plumage. The green head has a black lore and area around the eye and an indistinct black moustache. The mantle and underparts are also green, with a paler, more yellowy rump, while the tail shows yellow sides and a black tip. The wing shows a strong yellow edge at the front with grey coverts and tertials, contrasting with black wing tips, edged off-white.

A much drabber version of the male, female Greenfinches are, nevertheless, attractive in their way. The head, upperparts and most of the wings are a dull brownish-green, with the mantle showing some indistinct brownish streaks. The rest of the wing shows yellow along the outside and brownish tertials and coverts, grey on the male. The tail and rump are as on the male. The underparts are much as the upperparts, brownish-green, fading to greyish-white on the undertail.

A singing male Greenfinch will lay on a fairly impressive display with which to attract a mate. His twittering song is delivered both from a perched position and also in a delightful fluttering, butterfly-like display flight in which he circles the area to land back on a favoured treetop.

In the winter, Greenfinches become more inclined to visit gardens, particularly if a nutbag is hung up as a lure. The scene at a feeder can be filled with aggression and squabbling as birds try to oust each other from their positions.

Both males and females, when seen in flight, share a number of easily recognised field marks. Immediately obvious is the greenish rump and the bright yellow tail flashes, which contrast strongly with the deeply notched black tail fork. Also prominent in flight are the yellow blazes on the wing, which are slightly brighter on the male. The heavy bill and broad wings are other features to look for.

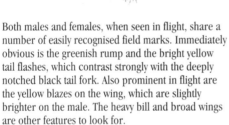

Young Greenfinches resemble the adult female, but show more streaking on the upperparts, including the rump, and are generally browner above and yellowish-grey below. The underparts are streaked brown.

GOLDFINCH

(11.5-12.5CM, 4³/₄ –5IN)

THE GOLDFINCH IS A SMALL, DISTINCTIVE FINCH which is common across the whole of Europe except for the very north of Scandinavia. It can be seen in a wide variety of habitats, from scrubby areas of wasteland to clifftop fringes, weedy fields, beaches, woodland fringes and gardens.

In the wintertime large flocks can be seen, on coasts especially, roaming between feeding sites. When disturbed, the splash of colour and clear, strong call make them utterly charming.

Quite a small finch, similar in size to the Siskin, the Goldfinch is a delicate bird with a largish pointed bill, rounded head, slim body and forked tail. The distinctive red, black, white, brown and gold plumage makes them quite unmistakable.

The 'face' of the Goldfinch is red, with black lores, crown and vertical ear-covert patch sandwiching white cheeks. The mantle and upper part of the rump are warm tawny-brown, and the wings are black with a prominent broad gold bar and white edges to the tertials. The uppertail coverts are whitish and the rest of the tail is black, with white fringes and central tip.

The underparts are white, except for tawny thumbprints on the shoulder and a brownish wash along the flanks. The undertail is black and white. The male (above) tends to show more red on the face, while the female (below) shows more grey on the wing. The pointed bill is whitish with a dark tip, the eye is black and the legs pinkish.

In flight, the black and gold on the wings are immediately obvious. Notice also the adults' distinctive face patterning, white rump patch and spots on the tail. From below the underwing looks two-tone (grey and white) and the white spots on the undertail are also prominent. The Goldfinch has a fast trilling song, very twittery, while the flight call is a clear, ringing "stickalit".

Goldfinches are especially fond of thistle heads and can perform with extraordinary agility when attempting to extract seeds.

Before their moult in the early autumn, juvenile Goldfinches are very dull compared with the adults. The head and upperparts are pale greyish-brown, lacking any red, black or white, the rump is buffy and the wings far duller. The underparts are buffy with dark streaks, particularly on the flanks. Some streaking may be apparent on the head. The bare parts are as those of the adults.

SISKIN
(11.5-12.5CM, 4³/₄ - 5IN)

THE SISKIN IS A CHARACTERISTIC, SMALL, RATHER DUMPY FINCH and is widespread throughout northern Europe. Its favoured habitats are conifer plantations, alder and birch woods and open countryside and, in the winter months, Siskins have utilised human influence by becoming regular visitors to gardens. They are particularly attracted to nutbags.

An extremely active finch, the Siskin proves to be a highly acrobatic bird when feeding, with flocks moving swiftly from tree to tree in search of food, and then often exploding from the treetops in a mass flock. A distinctive nasal twitter is heard from such flocks as they move.

Siskins have a distinctive slender, but sharply pointed, bill and a shortish but quite deeply forked tail. The plumage varies from yellowish-green to greyish-green above to white on the underparts, with variable amounts of streaking.

The male Siskin shows a distinctive mix of green, yellow and black plumage. The head has a black crown and bib, contrasting strongly with green cheeks and yellow on the rest of the face. The mantle is dark lime-green, with fine streaking, and the rump is a cleaner, fresher green. The wings are black with an obvious yellow wing bar. The underparts are clearly washed yellow, fading to white on the belly, with black streaking on the flanks.

The female Siskin is considerably duller than the male. The head and mantle are dull olive-green, streaked black. There is a broad, pale lime stripe about the eye, bringing some relief to the head pattern. The yellow on the wings is a little more subdued than on the male. The underparts show far more white on a female Siskin, with a dull green suffusion of the breast sides. The bold black streaks on the flanks are quite prominent.

◄ This is a typical view of Siskins in winter. A small feeding flock in an alder tree is performing truly amazing acrobatics. Their agility combined with non-stop activity as they hang suspended in every conceivable pose is a real joy to behold.

▲ The forked tail and long wings of the Siskin are markedly apparent when they fly. The black and yellow on the upper wing is particularly obvious, but also note the distinctive yellow tail sides.

▲ Juveniles show a distinctive pattern on the breast and flanks – the dark grey smudges contrasting with the otherwise clean whitish underparts. The upperpart coloration looks fresher than on female Siskins – a lighter grass-green tone.

► Rivalry between male Siskins can cause some quite spectacular squabbles, particularly where feeding is concerned. Aggression displays are commonplace on nut feeders, as an incoming male (right) is chastised severely by the dominant bird (left). Usually the dispute is merely a show of bravado, but occasionally the birds will tumble off and then recover, only to find that another bird has taken their place on the feeder.

LINNET
(13-14CM, 5-5½IN)

THE LINNET IS A COMPACT LITTLE FINCH, common throughout northern Europe. It is encountered, during the breeding season, in a wide variety of habitats from coast to heath, and outside the breeding season flocks are seen in stubble fields and meadows, as well as on roadside verges and in gardens. In many areas, however, it is an uncommon or even rare garden visitor.

As with most finches, family parties and winter flocks prove to be highly active, bounding between feeding areas and constantly 'twittering' at each other.

Linnets have a typical finch bill – slender and pointed – and a deeply forked tail. The plumage varies from the glorious pink and greys of a spring male to the delicate browns of the female.

The female (right) is more drab than the male. The head is streaked brown, with a greyish nape and creamy patches below the eye. The brown mantle is streaked blackish while the uppertail-coverts and rump are paler in tone, but still streaked. The wings, like the male's, are strongly patterned black and white. The underparts show a buffy-yellow breast, streaked brown, with the remainder white. The legs are more fleshy in colour than the male's.

A spring male (below) is a splendid sight, with a rich pink forehead contrasting markedly with the grey of the lores, crown, nape, ear-coverts and throat. The beady black eye is bordered by an indistinct creamy-white, very short, supercilium and spot below the 'cheek'. The mantle is a rich russet and the wings are marked strongly with black, brown and white. The breast shows strong pinkish-red patches, fading to ochre on the flanks. The rest of the underparts are dull white. The bill and legs are silvery-black.

◄ In flight, the familiar finch features – forked tail and long wings – are obvious. The russet of the male (left) contrasts strongly with the grey head and black wings, while the female (right) appears fairly nondescript, although the white on the shafts of the flight feathers and the tail is immediately obvious on both sexes.

▶ Juveniles appear similar in plumage detail to the female, but look significantly brighter. The head, mantle and wings are richer brown and the streaking is bolder. The breast is washed a gingery-brown, with streaks extending to the flanks. The bill appears quite silvery.

▼ The male Linnet in winter is a more subdued version of the rosy-tinted summer bird. The head is greyish, with darkish striations on the crown, while the whitish supercilium and stripe and 'spot' below the eye are slightly creamier than in the summer plumage. The mantle has many dark feather centres, looking quite scalloped compared with the plain-backed look of the breeding season. The underparts have quite heavy flank markings, brownish on a pale yellow wash. The breast is dashed with rusty-red. The bill tends to be more silvery than on the summer bird.

REDPOLL
(11.5-13CM, 4½-5IN)

THE REDPOLL IS A DISTINCTIVE FINCH commonly seen across almost all of western and northern Europe during various times of the year. It is found in gardens, parkland and various types of woodland. Redpolls come in various 'racial' guises and this accounts for their sometimes erratic distribution.

Redpolls that breed in Britain and Ireland are pretty sedentary, moving only locally in search of food, while birds on the near Continent move a little more, usually southeastwards. The Redpolls of the western European race are small and dark, while as you move further north-east the birds become a little larger and paler. Scandinavian birds are more inclined to move into eastern and western Europe, appearing in Britain in large numbers from time to time.

As with other tree-loving finches, Redpolls are agile, acrobatic and mobile, flocks often whizzing through overhead trees.

Female Redpolls differ from males in showing only a small amount of red on the cap and very little, if any, on the breast. The rump tends to be browner, too. The flanks are streakier and more buff-coloured than on the male. Like the male's, the bill is pale yellow, with a dark tip, and both sexes have black eyes, legs and feet.

Males of the British and Irish race show a small red cap on the forehead, black lores and chin contrasting with the buffy-brown cheeks, and slightly darker brown crown, nape and mantle which are flecked black. The rump is generally pale pink, the tail is dark greyish and the wings are brown-black with pale cream wing bars and feather edgings. White underparts show a pinky-red wash on the breast. The buff flanks are streaked finely with black.

▶In flight, notice the double buff wing bars, and paler rump. Listen out for the buzzing trill as they pass overhead.

▲ Birds of the Scandinavian race (female above, male right) appear larger and paler than the birds seen in Britain and Ireland. The head and upperparts appear quite frosty grey-brown, the wing bars and rump look whiter and the underparts appear paler, with buffy flanks and some streaking.

◀ Redpolls in winter appear even browner than they do in summer. Both sexes retain the red cap on the forehead, but the male's pink flush on the breast becomes quite indistinct, as shown by this male on the left. The rump colour of the male also becomes duller.

HAWFINCH
(16-17CM, 6-6½IN)

The Hawfinch is a shy, hulking brute of a finch which, although relatively uncommon, is found in many western European countries. The actual distribution of the Hawfinch is, as with several of the finch species, rather patchy. It breeds in England and Wales, eastwards through to Russia and southwest into the Mediterranean. In Scandinavia it is present only in the southernmost parts.

Hawfinches are rare visitors to gardens in Britain, favouring deciduous or mixed woodlands and parkland. In this country, chances of seeing the Hawfinch are greatly enhanced if you search suitable sites *early* in the morning, and look out for Hornbeam – the Hawfinch's favourite tree.

Their being so shy and wary makes study difficult, but Hawfinches are absolutely unmistakable with their large heads and huge bills. The striking brown, black, grey and white plumage makes them one of the most distinctive garden birds. Their call, in flight, is a loud, metallic 'zik-zik'.

The female Hawfinch (behind) is generally duller and more grey-brown than the male, particularly on the head, rump and underparts. There is less black around the face, and the primaries and secondaries are greyish, not black as on the male. On both sexes, the powerful bill is steel-blue in the summer, fading to brown in the winter. The pale brown eye has a large black pupil and the legs are dark pink.

The male (front) has a russet-toned head, except for the black feathering around the bill base and chin. The grey nape and collar merge into a rich mahogany mantle, fading into a russet rump and central tail.

The outer tail and tail tips are white. The wings show a broad buffy-white wing bar, contrasting with the bluish-black flight feathers. The underparts are pale honey-buff, fading to white on the undertail.

In flight, the large size enables good views of the short square tail and big head and bill. The russet in the head and rump will be easily seen, as will the grey nape. Look for the big white bar on the inner wing and the white shafts on the flight feathers, as well as the broad white tail band, clearly seen as the bird flies.

Juvenile Hawfinches are a greyish-brown, scaly version of the adults. The head is grey-brown, with a darker mantle, rump and tail. The upperparts are lightly scalloped blackish or dark grey, as are the buffy-white underparts. A hint of black may be evident around the bill base. The bill itself is off-white to greyish, the eye is yellowy-white and the legs are flesh-pink.

YELLOWHAMMER
(16-17CM, 6-6½IN)

The Yellowhammer is a large slim-looking bunting which is easily seen throughout Europe. It is found in open country, large gardens, farmland and hedgerows. In winter Yellowhammers can often be found in mixed flocks of finches and buntings.

Yellowhammers are elongated birds with long wings and a longish notched tail. The plumage of the male is especially distinctive, particularly in the early part of the breeding season, when its head is a glowing sulphurous yellow, with a variable amount of dark olive streaking on the centre of the crown, and a blackish line behind the eye that curves downward across the cheek. The rump is even brighter and is unstreaked. Even the drab female is fairly unmistakable.

▶ A male Yellowhammer is a real sight to behold! The nape is a rich olive, and the mantle is a striking chestnut, streaked heavily with dark brown. The constantly flicked tail is blackish with prominent white outer tail feathers. The wings show a neat chestnut, black and white patterning. The underparts are variable in pattern, but generally show a yellow upper breast and a rich chestnut lower breast patch which then becomes streaked on the flanks. The belly is variable – from pale yellow to white, edged black. The bill is silvery-grey, the eye black and beady, while the legs and feet are fleshy-pink.

▼ The female's head is yellowish, with a finely streaked crown, dark-bordered greyish cheeks and moustache. The throat is dull yellow. The nape is brownish, with fine dark streaks, and the mantle is similar in tone but shows heavy blackish streaks. The rump, tail and bare parts are as the male's, as are the wings but not so chestnut. The underparts are streaked brown. The belly and rear flanks tend to be pale yellow or white.

Yellowhammers flock in winter and become regular
visitors, with other buntings and finches, to large
gardens to feed on seed and grain. They are also
common, during winter months, in stubble fields and
farmyards. After moulting from juvenile plumage in
July to October, first-winter birds resemble adult
females, but can be separated. Young males are
brighter yellow on the head and underparts, and the
chestnut of the wings, flanks and rump is brighter.
Young females show a dark-streaked brownish crown,
a dull olive-yellow head, buff throat and pale belly.

A male in flight shows an obvious yellow head, dark
wings, streaked mantle, plain chestnut rump, dark
tail and white outer tail feathers. Notice how long the
tail is, and look for the deep notch.

A juvenile looks like an adult female, but shows more
streaks on the head, streaked breast and flanks (both
washed yellow), and is generally much duller even
than a female. The rufous on the rump is dull and
faintly streaked.

THE REED BUNTING IS AN ACTIVE BIRD which is now becoming a more regular visitor to people's gardens, particularly in the winter. It is common throughout northern Europe, and frequents not only marshes and reedbeds, but also hedgerows, bushes and, particularly in winter, farmland.

As with the Yellowhammer, Reed Buntings often associate with other small seed-eating passerines during the autumn and winter, feeding in stubble fields, gardens, winter-wheat fields or at special feeding stations. They have a small, stubby seed-eating bill which is convex in shape, with a black eye and dull flesh-coloured legs and feet.

Both sexes have distinctive breeding and winter plumages, the male being particularly striking and in summer plumage quite distinctive. The mantle, rump and bare parts are similar in both sexes, although the female's tail can sometimes be browner than the male's. Both sexes are slimmer, less bulky and shorter-tailed than Yellowhammers.

This spring male is showing its jet-black head with white moustache and neck collar. The black extends from the throat onto the centre of the upper breast. The mantle is blackish, with rufous feather edges and broad yellow tramlines extending to the greyish rump, which is lightly streaked with black. The uppertail is dark brown-black with bold white outermost feathers. The underparts are white, with fine black streaks on the flanks. Males sing from exposed positions, the song being an unforgettable slow wheeze, of four or five notes.

Females lack the male's black head, but are otherwise similar. The forehead and crown are chocolate-brown, streaked with black. The cheeks are bordered by a white supercilium and moustache and the throat is white. The underparts are white, streaked on the breast and flanks.

▶ Juvenile Reed Buntings resemble the adult female, but have a darker head, are more buff below and are well streaked. The crown and cheeks are brown, with a brownish supercilium. The collar is greyish, but the remaining upperparts are much as the adult's. The underparts are washed entirely buff-brown, with a heavily streaked breast, less well marked on the flanks. The undertail is off-white. The bare parts are as on adults.

◀ A winter female shows a greyer head than a 'summer' bird, with a yellow wash on the supercilium and throat. The markings on the breast and flanks become more pronounced, with longer streaks and brown along the flanks.

▼ The head pattern of the male Reed Bunting in winter becomes a little tainted by rufous tips on the feathers, particularly around the lores, supercilium, ear-coverts and throat. The white collar is still visible, and there is always a hint of the black throat. The wings can appear rather chestnut. The underparts show some indistinct grey flecks on the flank.

USEFUL ADDRESSES
AND FURTHER READING

USEFUL ADDRESSES

BTO (British Trust for Ornithology)
The National Centre for Ornithology, The
Nunnery, Thetford, Norfolk, IP24 2PU.
Tel: 01842 750050.
The BTO organises a range of surveys, designed to monitor changes in the population of Britain's birds. These include the Garden BirdWatch, in which the owners of more than three thousand gardens all over the country record the birds there. If you would like to participate, write to the BTO for further details. Annual membership of the BTO costs £19 for adults, and £26 for the whole family. Members receive a bi-monthly magazine, *BTO News*.

RSPB (Royal Society for the Protection of Birds)
The Lodge, Sandy, Bedfordshire, SG19 2DL.
Tel: 01767 680551.
The RSPB is Britain's leading bird conservation organization, with almost a million members. It can provide a wealth of information, advice and practical help on attracting birds to your garden. The RSPB also has a mail-order service, selling approved bird-tables, nestboxes, bird-feeders and many other items of equipment. Annual membership costs £20 for adults, and £28 for the whole family. Members receive four copies of *Birds* magazine each year.

YOC (Young Ornithologists' Club)
The Lodge, Sandy, Bedfordshire, SG19 2DL.
Tel: 01767 680551.
The YOC is the junior arm of the RSPB, and has more than 125,000 members under 16. Annual membership is £7 for one child, or £9 for all the children in a family. Members receive a bi-monthly magazine, *Bird Life,* as well as a host of other benefits such as birdwatching trips.

CJ Wildbird Foods Ltd
The Rea, Upton Magna, Shrewsbury, SYA 4UB.
Tel: 01743 709545.
CJ Wildbird Foods are a leading supplier of approved birdfood of all kinds, as well as nestboxes, feeding equipment, birdtables and much more. They produce a free guide and regular supplement detailing a huge choice of bird food and accessories.

FURTHER READING

Books:

The Bird Table Book Tony Soper
(David and Charles, 1992)

Creating a Wildlife Garden Bob and Liz Gibbons
(Hamlyn, 1988)

The Complete Book of British Birds
(AA and RSPB, 1988)

Feed the Birds Tony Soper
(David and Charles, 1991)

The Garden Bird Book David Glue
(Macmillan, 1982)

How to make a Wildlife Garden Chris Baines
(Elm Tree, 1985)

The New Atlas of Breeding Birds in Britain and Ireland: 1988-1991
(T. & A.D. Poyser, 1993)

The RSPB Birdfeeder Handbook Robert Burton
(Dorling Kindersley, 1991)

A Photographic Guide to Birds of Britain and Europe Paul Sterry and Jim Flegg
(New Holland, 1995)

Magazines:

Birdwatch
Available monthly from larger newsagents, or by subscription from:
Birdwatch, 1 Northumberland Park I.E., 76-78 Willoughby Lane, London N17 0SN.

Birdwatching
Available monthly from larger newsagents, or by subscription from:
Birdwatching Subscriptions, Tower Publishing Services Ltd, Tower House, Sovereign Park, Market Harborough, Leics LE16 9EF.

BBC Wildlife magazine
Available from most newsagents, or by subscription from:
BBC Wildlife Subscriptions, PO Box 425, Woking, Surrey GU21 1GP.

INDEX

Note: Numbers in bold type refer to main entries and those in italics refer to illustrations.